COMPLEXITY THEORY AND PROJECT MANAGEMENT

COMPLEXITY THEORY AND PROJECT MANAGEMENT

Wanda Curlee and Robert L. Gordon

John Wiley & Sons, Inc.

This book is printed on acid-free paper. ♾

Published by John Wiley & Sons, Inc., Hoboken, New Jersey
Published simultaneously in Canada

Library of Congress Cataloging-in-Publication Data:

Curlee, Wanda.
Complexity theory and project management / Wanda Curlee, Robert L. Gordon.
p. cm.
Includes index.
ISBN 978-0-470-54596-6 (cloth); ISBN 978-0-470-76972-0 (ebk);
ISBN 978-0-470-76973-7 (ebk); ISBN 978-0-470-76974-4 (ebk)
1. Project management. 2. Complexity (Philosophy) I. Gordon, Robert L. II. Title.
HD69.P75C87 2010
658.4′040117–dc22

2010018383

Printed in the United States of America

10 9 8 7 6 5 4 3 2 1

Contents

Preface

Organizations and projects are changing...

There is at present a great deal of interest in understanding projects in terms of complexity theory, self organization and emergence. The image of project manager's surfing on the edge of chaos is prevalent in practitioner conferences. I've used it myself to talk about the changing requirements for education for advanced project managers. As we acknowledge that projects are less structured and linear than we would like, we must realize that today's methods may not work on tomorrow's projects.

There is also increasing pressures to work in globally networked teams. When you can't see, touch, or have a beer with your teammate, how do you develop the trusting relationships necessary to deliver exceptional team results?

The people making up project teams (collocated or virtual) are more diverse—multi cultural, multi generational, multi skilled. Younger workers with possibly lower concentration spans and much higher aptitude and inclination to use technology intermingle with older workers used to set processes and procedures. "New" Americans and foreign colleagues work on projects with "old" multi cultural residents. All of these demographic changes are sure to influence how we work on projects. But how?

These are the challenges that academics around the world have been interested in for years. These are also the challenges that PMI's Research Membership Advisory Group is charged with encouraging research on. Almost 10 years ago when I was sitting on this committee I met a steely eyed woman by the name of Wanda Curlee. It quickly became clear that this woman both defied categorization having practised project management as a consultant and inside organizations and holding a doctorate degree in organizational leadership, and was as passionately intrigued by these challenges as anyone I have ever met. One of the good things about this committee is that it requires you to meet several times a year in exotic locations (well sometimes exotic) and after the business of the meetings, this group gets a chance to talk through the problems and challenges of managing projects in an ever changing world – and in particular of bringing research to bear on practice. I have had the privilege of many such conversations with Wanda over the years.

These are also the challenges that Drs Wanda Curlee and Robert Gordon tackle in this book. More than that, they integrate discussions of complexity with understanding of current project management standards and weave a set of tools and tips together to help the practicing project manager understand how these trends are likely to impact his/her practise. They provide relevant case examples and suggest practical tools and techniques. They bring to bear both the art and science of project management to explore projects as complex, emergent systems in this timely book.

Organizations the world over are grappling with the ideas and challenges discussed in this book. Just this week I personally spoke to the CIO of a major educational institution, the PMO of a small, 3B$ oil and gas company, and the person responsible for project management development at one of the world's largest consulting companies. All of these individuals were looking for guidance on how to develop organizational and individual project management competency to ensure that they get value for dollars invested in project management. They call me to hear what practical insights we learned from studying 65 organizations all around the world. All three of these organizations are dealing with situations where complexity and chaos are much closer to reality than linear, planned project execution. Ideas from these pages could help them today, and you tomorrow.

I encourage all experienced students of project management to read this book and apply as many ideas from it to your practise as you can. Come on in, the waters fine and surfing is a skill we all need to develop.

Janice Thomas, PhD
Redwood Meadows, Alberta, Canada

Director, Project Management Research Institute
Professor, Athabasca University, Canada
Leader, Researching the Value of Project Management Project
PMI Research Achievement honoree 2010
Winner of the 2010 PMI Research Achievement Award

Acknowledgments

This project started years ago with an idea that was bantered around in project management circles. It was not an easy project to define nor was it one that was easily researched. The material that directly related to the use of complexity theory and project management was sparse and in many cases it required searching for information that might not have been called complexity theory but was exactly that. In a few instances authors might credit chaos theory, but most of the time it required sifting through the body of knowledge of successful projects and organizations and then delve into the material to see if it met the criteria of complexity.

Just like the idea of this book first started as an idea in project management circles, the writing of every great work starts as an idea that slowly grows up until it takes on a life of its own. This book is the union between project management and complexity theory. The marriage of those two bodies of knowledge was not a simple task, and in the end the book required a team of individuals who were willing to contribute a part of each of them in order to achieve the work that is before you.

There is a saying that claims that it takes a tribe to raise a child. In this case, this saying is absolutely correct, as the child in question is the book that is before you. And the tribe is the loose confederation of people that all put their time, sweat and tears to get this project completed.

We would first like to acknowledge the tribe of photographers that helped bring this book to life. Spencer Ludgate, Nicole Burdsall, and Helle Girey all contributed pictures that helped bring home some of the ideas and concepts of complexity. Each of them brought pictures from the desert, to the ocean, and all the way from Africa, these illustrations helped offer representations of complexity that are seen in the natural world around us.

Second, there was the tribe of individuals that assisted with editing and indexing of this book. Our primary initial editor was Roseann Kruger. We would like to offer special thanks for her efforts to read every page of this book aloud in order to catch our errors, omissions and mistakes. It must be noted that the entire time that she was assisting us with this book; she was undergoing treatment for cancer. We are eternally grateful for her determination to see the project through and for her uncompromising assistance when it was needed most. We also want to offer our heartfelt thanks to Cassandra Corl for her assistance with the indexing of this book. She assisted on short notice and just as fast as a rainbow comes and goes, she was there to get the job done.

Both of our family tribes spent many hours wondering why we continued to spend days on the computer, instead of precious family time with them. We thank our families for being patient, understanding, and most of all for the encouragement to continue. Our time of adding to family complexity has quieted, we now acquiesce to others.

We would also like to thank that team at Wiley, all of whom were as professional as they come; whose efforts and feedback made this book a reality. The communication, support from the group at Wiley made certain that this book would be the best in the field of complexity and project management. In the end, it is clear that a tribe built this book and there is no doubt that the work of a community that makes a difference in society.

All the best,
Wanda Curlee & Robert Gordon

Introduction

Complexity theory is based upon the management belief that total order does not allow for enough flexibility to address every possible human interaction or situation. The problem is that people are inherently skeptical of less order and flexibility because there appears to be less control. A recent example of the working of complexity theory can be seen in the investigation of air traffic controllers after the 9-11 tragedy. Once it was determined that terrorists were using airplanes to attack buildings in the United States, it became a matter of national security to have every plane in the airspace of the United States land at the nearest airport. There was no procedure or process in place to allow this to happen.

Researchers were interested to see if a procedure or process could be developed to address such a widespread domestic crisis. The researchers studied the data and examined how each set of air traffic controllers managed the situation. In the end, the study concluded that the best way to handle such a crisis would be to allow each region to dynamically manage the situation. In other words, the creation of a procedure or process to handle such a situation would encumber the process and slow down the ultimate goal. This was a massive eye-opener for project managers everywhere because it was the first time that there was a multi-location study for a single industry whose results were not the creation of a linear solution to what would be considered a linear problem. It should also have been an awakening for project managers because it brought to light the inherent flaw in the underlying assumption that there is always one right solution or procedure to a problem.

For management theorists, this was a win for complexity theory because it showed that a complicated and complex project can be more successful utilizing complexity theory rather than looking for the single management solution to a seemingly linear problem. Complexity theory is a new, untapped reservoir of potential in the management field. Experts agree that complexity theory can apply to complex, virtual projects;

however, there is little material that would help the practicing project manager. Complexity theory has become a recognized area of project management and the *Project Management Body of Knowledge (PMBOK®)* Guide should address this area in the future. Complexity theory can be successfully applied to a complex, virtual project and so should become part of the *PMBOK Guide®*.

There is no handbook of practical and successful strategies that applies complexity theory to projects. Given the rate of failure of projects when

Figure I.1 Notice how the clouds reflect upon the building but not upon the sky. This is an example of how everyday images might offer a different and complex perspective.

compared to timeline and budget, it would be extremely valuable to offer practical complexity strategies that a project manager can deploy in order to improve his or her success rate.

Specifically, this book is broken down into five major sections. Part I outlines the theory of complexity and addresses current deficiencies in the existing body of knowledge about project management, and introduces ways that complexity can address these deficiencies. Part II addresses successful strategies that deploy complexity theory in order to make projects more successful. Part III presents case studies and details regarding how complexity theory has been applied successfully in other organizations. Part IV offers building blocks for project managers on how to create communities within their organization to support complexity theory. Part V reviews and summarizes the findings and reviews the future of the application of complexity theory. Complexity will continue to grow in importance in the project management field as more organizations understand how to apply these ideas to projects. Ultimately, the goal of any book on project management should be to improve the field of project management. Hopefully the knowledge set forth in this book will help project managers worldwide to become more productive and successful while maintaining their sanity.

Part I

Complexity Theory

Part I provides an introduction to complexity theory and virtual projects. The first part presents a traditional introduction to the history of complexity history and how complexity theory is used in the business world today. This is followed by a review of major project management associations, their discussions and accommodations of complexity in standards and journals. Virtual projects and leadership are introduced with a short orientation on complexity and how it relates to the distributed environment. Part I ends with examples of successful (the champions) and unsuccessful (the mutts) virtual projects, and how applying complexity increased the likelihood of a champion.

A PRACTITIONER'S EXPLANATION OF COMPLEXITY THEORY

Seasoned project managers realize that all parts of the projects cannot be controlled; nor would they want to have full control of the project. They realize the creativity occurs on the fringes of complexity or chaos. Those teams that appear to be in total chaos may be doing the best work for the project; those work teams that have a catastrophic failure of some sort and are allowed to resolve it on their own do so more quickly and efficiently. This appears to go contrary to standards suggested for project managers.

In the academic sense, complexity theory does go against the common teaching of controlling all aspects of the project. Mounting evidence—both anecdotal and academic—demonstrates that the traditional method just is not working. The Standish Chaos Report (2009) clearly shows that software projects continue to fail more than 65 percent of the time. Yet it is hard to believe the project managers are becoming less competent.

The Standish Report gives the same reasons from report to report. Incompetent project managers are never in the top ten reasons. There is no doubt the reasons are correct but there must be some underlying factor as well. The suggestion is that the seasoned project manager knows how much to control the chaos, what absolutely needs to be controlled, and what can be left to chaos.

This wisdom does not come by chance or happenstance. The project manager must completely understand the theory and implementation of project management. The project manager *must* understand the integration of the processes and must have the self-confidence of his or her ability as a leader. Once these are in place, then the project manager is ready to embrace the ability to allow chaos on the project.

Chapter 1

Introduction to Complexity Theory

Figure 1.1 The Joshua tree appears to be a hand reaching toward the sky, just as a project manager must reach toward new ideas in order to be successful in the future.

INTRODUCTION TO COMPLEXITY THEORY

Complexity theory evolved from chaos theory, and there has been written evidence of the theory in the scientific community since the 1800s. Relatively speaking, these are very new sciences (Singh & Singh, 2002). Parts of complexity theory are even considered mainstream. Two aspects are commonly referenced in the mainstream media are the butterfly effect and six degrees of separation.

Complexity theory has been alive and well in the areas of math and sciences. Complexity has slowly moved into the areas of social sciences and is now making a step into the world of business (Byrne, 1998). From the social networking perspective (six degrees of separation), complexity theory has made a quantum leap, as discussed in more detail in later chapters. The butterfly effect, if allowed, can be very effective for the virtual project as well. This, too, will be discussed in later chapters.

Complexity theory acknowledges that humans by nature when living or working together are an open system (Byrne, 1998: Hass, 2009). What makes complexity theory different than the traditional open systems theory is that the theory acknowledges that there are parts of the system that cannot be explained but acknowledges that there is normalcy in the randomness. Human beings like to break down the system into its smallest part to explain the whole. Western thought seems content to understand the universe as a series of discreet system rather than a holistic interconnected system. Using this approach to examine, for example, how a single ant works independently, does not explain the dynamics of the colony. Or explaining how the human heart works does not explain the interrelationship of the glands, brain, heart, blood, and so forth, and what happens if one part is out of control. In other words, how will one part of the body compensate for others?

Hence, if complexity is introduced into the system, should it be managed? Think about the ants that are away from the colony on a mission. If an unexpected situation is introduced into the army of ants on the mission or those in the colony, each section of ants begins to react no matter what the situation. Some would say it is instinct. Maybe it is. Without any central authority or predefined processes, the ants resume to their goal. Some of the ants are more central to the goal, while others are more tangential to the goal, but all must perform as a team to ensure the queen's survival or the heirs' survival will fail. Is a project any different?

> Practical Tip: Try to understand that all systems are connected. By understanding these interconnections, new understanding can be achieved. Explaining matters as discrete silos does not explain the bigger picture. Understanding the role of the quarterback in football does not explain how football is played.

HISTORY OF CHAOS THEORY

For many years, chaos theory was a hobby among mathematicians and scientists. It was not taken seriously because a real world application could not be envisioned. Edward Lorenz, a meteorologist, is credited as the first person to delve into chaos theory. In fact, he is the person credited with the *butterfly effect* theory. Simply stated, this theory holds that when a butterfly flaps its wings in another part of the world, say South America, this creates a minute disturbance in the atmosphere. This minute disturbance may have a drastic change on the weather conditions in North America. It may create a hurricane or it may prevent a hurricane where one should have started.

What did Lorenz find from all his experiments in the realm of chaos? Many things that do not appear to have any order actually do. Scientists would have called this noise or randomness but it is indeed chaos theory at work. Lorenz's attractor equation was able to clearly demonstrate order in the world of chaos. He found that the atmosphere never reaches a state of equilibrium. Hence it is always in a state of chaos. When plotting the atmospheric conditions, it always plotted as butterfly wings or owl eyes (Wheatley, 2000). So there was order in the randomness (Figure 1.2). In essence, the atmosphere disturbances were drawn to areas or attractors. Thus, it appeared as order in what previously was thought to be randomness. As a meteorologist, he did not have credibility in the mathematical and physics communities. For many years, his work went unnoticed.

With the introduction of high-speed computers and the observation of chaos in the areas of fluids, semiconductors, and other hard-core science areas, this area of science flourished. Chaos theory remained in the area of hard-core sciences for many years, since it is often difficult to make the leap from one area of academia to another. It would be another ten years or more before enterprising project managers and academics in the area of project management would see the correlation.

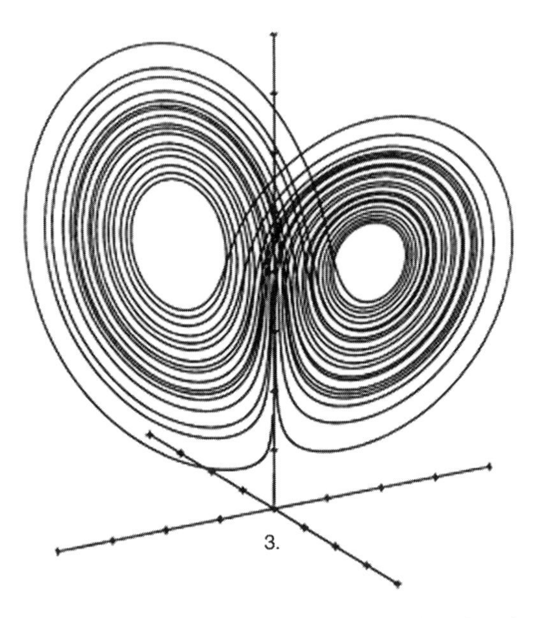

Figure 1.2 Lorenz's Butterfly Attractor Model (Wheatley, 2000)

Chaos theory would still have to go through some changes. It would shift from chaos theory to complexity theory. The theory would shift into various areas of study as it matured. This is no different than what is seen with mature professions such as medicine. Today, there are many different specialties including cardiology, neurology, ophthalmology, and so forth. The human body is probably one of the most complex systems known; even skilled surgeons do not know exactly how each of us will be in surgery. Even if great care is taken prior to surgery, there are no guarantees that the procedure will be successful and without unforeseen complications.

As the area of complexity theory matures, the same would be expected to occur in project management. In fact, the Project Management Institute^{TM} (PMI) has expanded its certification program to include risk and schedule management. Risk is becoming a more widely accepted aspect of both project management and complexity. Although the concept of contingency has existed, this concept supposes that there are controllable unknowns. Complexity accepts that there are simply unknowns and the best manner to handle these would be to have a flexible process rather than a rigid contingency (Weaver, 2007a). This step is the first in accepting that complexity exists in projects, and one can be certain that the future of project management will be more inclusive of this kind of training.

HISTORY OF COMPLEXITY THEORY

Complexity theory grew from chaos theory. The theory works on the notion that a system should not be broken down into fundamental parts to understand the whole. This should not be confused with the theory and studies of self-organizing teams that are seen in the business world.

Complexity theory states that critically interacting components self-organize to form potentially evolving structures exhibiting a hierarchy of emergent system properties. (Lucas, 2006)

Chaos theory offers a view of the universe that everything is not as orderly as once thought. There are many seemingly immutable laws, such as the speed of light, the forward movement of time, and the force of gravity; there seem to be other natural phenomenon that defy this kind of explanation. In some cases, one must understand the entire system in order to understand how all the parts interact. Just as one might understand that the speed of light is 186,000 miles per second, there are still times that light seems to act differently, such as when interacting with a black hole. Time, light, and gravity are seemingly different natural laws, yet there is still a systemic relationship between the three.

As an entity approaches the speed of light, the movement of time apparently slows down, and as light interacts with the intense gravity of a black hole, light seems to move unnaturally. Again, this interaction between three apparently immutable natural laws fundamentally supports the necessity for chaos theory, and by extension, complexity theory.

From another angle, there is simply too much interaction among any natural systems. Weather, which is another chaotic system, is too complicated to fully understand. There are simply too many parts that interact and our understandings of these interactions are based upon observation rather than upon a mathematical model. There are certain truths, such as there cannot be rain without clouds, but there are many instances where this kind of simplistic modeling does not describe the system (Weaver, 2007b). Fundamentally, many natural and human systems are incapable of being described by a mathematical model. Few human systems can be properly described and predicted through single variable mathematical modeling. Too often individuals believe that certain systems are predictable based upon mathematical modeling. This becomes a major stumbling block toward the acceptance of complexity theory. The reality is that with most human-based systems, there are too many variables to predict the results. Just as there is no perfect way to forecast the weather or to predict whether an individual's favorite soccer team will be victorious in its next game, most human social systems are similarly unpredictable. What often befuddles individuals is that statistical modeling is representative of groups and relationships and not of behavior. Hence, statistical modeling is helpful in understanding groups and relationships but does not help us understand individuals and social systems. This [illegible] leadership crux because as a leader, one needs to utilize all the available project management information in order to be successful, but it does not offer a clear path. Simply repeating a known strategy and hoping that it will be successful is not the best course of action.

As complexity theory is maturing, scientists, whether in hard-core science or in business or behavioral aspects, have come to realize that complex systems cannot be viewed broken apart to understand the whole. They have realized the importance of applying complexity theory to business. Think of the old argument of nature versus nurture, or of identical twins separated at birth. Science for many years has not been able to prove what is nature and what is nurture or if it is interrelated. Why do some individuals faced with almost the same familial situations end up with radically different reactions or personalities? While it cannot all be contributed to complexity theory, it would be reasonable to apply complexity theory to the phenomenon.

COMPLEXITY THEORY IN USE TODAY

Several elements of the manifestation of complexity theory are already in use. One of those elements is transformational leadership. In order to understand how transformational leadership differs from leadership of the past, it is necessary to understand the evolution of leadership theory. The history of leadership can be viewed as a continuum of eras leading up to leadership in the virtual environment. Just as society has changed, leadership has changed in a manner to reflect these sociological changes. Some research has tried to categorize leadership into different schools of thought, such as *old school* and *new school*; this kind of division creates a feeling that there was a split in thinking. This kind of division is based upon one simple rule of leadership theory of the past and of the present.

In the past, there was an attempt to find the one right way to lead people. This assumption was rooted in management theory for a very long time and whenever a theory came forward that did not cover all the possibilities, it would be rejected and then the next management theory would come forward. Over time more management theories came forward, and often they would reflect society, technology and business. As leadership evolved, more ideas were presented and rejected. The interesting aspect about this is that over time, many of these ideas became integrated into modern thought.

[illegible]e is more of an evolution of thought that reflects society, [illegible]e, and even business. To this end, leadership has evolved from one stage to the next and in each step of the process certain elements have been carried along to the next. Transformational leadership and its characteristics of mentorship and learning adapt well to complexity within a project. Understanding that complexity theory, like mentoring, is a journey and not a destination is to better understand the concept (Huang & Lynch, 1995). It is not a finite set of skills but a constantly changing opportunity. Just as one can never enter into the same river, complexity is about learning to accept certain unknowns with flexibility and grace.

> Practical Tip: *For the complex project, complexity theory suggests that costs should be forecasted following each butterfly effect episode. Do not impose strict cost forecasting as it may limit the effectiveness of the process (Overman & Loraine, 1994).*

Chapter Summary

Complexity theory is about harnessing chaos in a manner that allows the project manager to increase his or her team's effectiveness by allowing a certain degree of individuality to move a project forward. Often permitting the random walk of the determined individual allows a certain level of creativity to become successful. An effective team can be more effective than an individual; allowing an individual to plow forward can often drive the team further and faster. Complexity is the manifestation of empowering and delegating tasks to allow individuality to support the hive.

The *butterfly effect* is the understanding that all forces are connected. When a project is moving forward, it is best to try to put all the forces working in the same direction. Just as the flapping wings of a butterfly in Japan can be a contributing force to the creation of a hurricane in Florida, understanding that even a small impact can have a great effect when magnified over time and distance. Hence, a leader who can motivate each individual can assist in creating a controlled hurricane that can achieve complex tasks. Too often people do not realize that even small contributions can build to create something larger than their individual parts, and so the small contributions are ignored. The more that virtual project managers can harness this kind of organization, the more effective they will become.

Lines of communication are critically i [illegible] as it is the representation of command and control. It also represents the fluidity of how information is exchanged. By understanding how this impacts a project, a certain level of strength develops from this form of communication. Just as the arteries carry blood throughout the body, the lines of communication carry information through the project. Taking this metaphor one step further, one can see that any blockage in an artery will have catastrophic effects, just as communication blockage would have a catastrophic effect in communication.

A project manager must prepare for change within a project and must retain a level of connection with contacts and leads. Complexity theory is more than just being a laissez-faire leader who issues discretionary orders; it is about creating a solid purpose that the individuals can swarm toward. Instilling the purpose is at the root of success of complexity theory. To this end, the project leader must offer support of this process in order to have it continue in the future. Complexity theory will move more projects in the future because as these types of projects become more successful, more organizations will understand the greater efficiency of these organizations.

CASE STUDY: LOOKING FOR COMPLEXITY WITHIN A PROJECT TEAM

In order to be successful at leveraging complexity in a project, a project manager must be able to identify circumstances that already could be leveraging complexity. Following are four short descriptions of situations that could be leveraging complexity. Consider each case and how it relates or does not relate to complexity.

Situation 1: The project team that you have inherited is utilizing a social networking site in order to report progress to other stakeholders. The team is fairly consistent about reporting progress on the site, but the site has not been communicated to everyone. Is this situation one that is leveraging complexity? Why or why not?

Situation 2: The project team sends a weekly newsletter to all stakeholders to report progress. This newsletter is the primary communication to all the stakeholders and is updated regularly. The information sent often generates many questions from different people that require time to be addressed. Is this situation one that is leveraging complexity? Why or why not?

Situation 3: The project team is in flux since the team has decided to reconfigure its structure in order to better handle the project. Some people are unhappy with the changes; a lot of processes have resulted from the changes. Certain impediments have been removed, certain procedures that were unproductive have been changed or modified, and communication is flowing better to all the stakeholders. Is this situation one that is leveraging complexity? Why or why not?

Situation 4: The project team is behind schedule and the project will probably end over budget. Information has been passed among the team regarding the delays and challenges; many team members are already concerned about how the project will end. Individuals have started to be less communicative and fewer updates have been sent out about the project. The overall team feeling is that the project will end poorly, but there has been no formal organizational announcement that things may not be right. The project lead is reluctant to report the possible situation because the most recent milestones have been achieved. Is this situation one that is leveraging complexity? Why or why not?

CASE STUDY REVIEW

All four of these situations are leveraging complexity. What is interesting is the first three are positive displays of complexity in action within an organization. The last scenario is actually a very negative situation with complexity in action. Complexity can assist a project manager in becoming more successful; it can also offer additional problems for a project. In many cases when a project goes terribly wrong, it is often due to the negative elements of complexity.

In the first three situations, there is good communication, good handling of change, and an understanding of the transformational nature of the project. These three scenarios offer creative and interesting manners to handle a project while still maintaining some control and order.

The last situation is a case where negative complexity is keeping the project from getting the help or assistance that it needs. The project team in the last scenario is becoming defensive in order to keep the underlying problems of a project a secret. Just as complexity is about offering communication, change, and leadership, these elements exist in the last scenario but the outcome is to keep everyone in the dark about the project. This is a classic case of trying to hide the project from sight in the hopes that things will get better. There is often an element of delay and it may go away in the mindset of the last project. The efforts of the project team are now pointed inward in order to keep the secret and to take the vow of silence. Even the project leader is behaving in this manner. Rather than trying to seek help or to start the process of determining the root cause of the project failure, the project manager has decided to hide.

Project managers must understand that complexity can operate both ways—as a positive to a project or, if improperly applied, as a negative to a project. Change follows in the same manner, as it can be seen as either a positive to a project or as a negative (Brown & Eisenhardt, 1997). Complexity is perceived in exactly the same way by project managers. Just as it offers new ideas and new manners to correct problems, it can also be used to hide problems and suppress change. Understanding that both situations are possible is critical in achieving future success. In order to make a difference, one needs to understand that complexity can yield good things, but it also can be used by others to obfuscate a project. Just as a magician cleverly creates a distraction when the deception is about to take place in the performance, complexity can be used in a manner that is no different. A project manager who is weak or may be less than ethical may use complexity as a crutch. Complexity in a project is not easy to oversee. It is difficult and takes time. The weak project manager may use complexity as a cover for his or her lack of skill.

REFERENCES

Brown, S., & Eisenhardt, K. (1997). The art of continuous change: Linking complexity theory. *Administrative Science Quarterly*, *42*(1), 1. Retrieved August 21, 2002, from Business Source Premier.

Hass, K. (2009). Managing complex projects: A new model. Vienna, VA: Management Concepts.

Huang, C. & Lunch, J. (1995). Mentoring: The Tao of giving and receiving wisdom. San Francisco, CA: Harper

Lucas (2009). Quantifying complexity website. www.calresco.org/lucas/quantify.htm

Overman, E. & D. Loraine. (1994). Information for control: Another management proverb. *Public Administration Review*, *54*(2), 193–196.

Pievani, T. & Varchetta, G, (2005). The strategies of uniqueness: Complexity, evolution, and creativity in the new management theories... or, in other words, what is the connection between an immune system network and a corporation. *World Futures*, *61*. Milan, Italy: Routledge Taylor Francis Group.

Project Management Institute (Ed.). (2008). *A Guide to the Project Management Body of Knowledge—Fourth Edition.* Newtown Square, PA: PMI.

Project Management Institute (Ed.). (2009). *Practice Standard Project Risk Management.* Newtown Square, PA: Project Management Institute.

Standish Chaos Report. (2009). Standish Group.

Singh, H., & Singh, A. (2002). Principles of complexity and chaos theory in project execution: A new approach to management. *Cost Engineering*, *44*(12), 23.

Weaver, P. (2007a). Risk management and complexity theory: The human dimension of risk. 2007 *PMOZ Conference Proceedings*.

Weaver, P. (2007b). A simple view of complexity in project management. 2007 *PMOZ Conference* Keynote address.

Wheatley, M. (2000). *Leadership and the new science: Discovering order in a chaotic world* (2nd ed.). San Francisco: Berrett-Koehler Publishers.

Chapter 2

Going beyond the *Project Management Body of Knowledge* (*PMBOK*®) *Guide*

Figure 2.1 The rock formation creates a puzzle of stones that must fit together to create a solid standard—just as the body of knowledge of project management should be codified in the *PMBOK*®.

The Guide to the Project Manager's Body of Knowledge (*PMBOK® Guide*) (PMI, 2008) is an essential document for any project manager. The purpose of the guide is to offer a standard for project managers in order to assist them with the handling of projects. Because projects can be complex—such having to manage teams of people spread over the globe—the *PMBOK® Guide* is often silent. Since a project manager has many options to solve a project and since project management is a growing field, the *PMBOK® Guide* cannot be reasonably expected to have all the answers. Given this quandary, the modern project manager must examine other material in order to find appropriate solutions for complex projects. Even the *PMBOK® Guide* itself in Section 1.1 advises that just because a topic (such as complexity theory) is not addressed in the guide as a project management topic, does not mean the material is unimportant (PMI, 2008).

The *PMBOK® Guide* cites three different reasons for omission of a topic like complexity. First, the guide states that it could be addressed in an existing section and is just a subset of a larger topic (PMI, 2008). Some individuals perceive complexity as part of the general project management process of delegation; this is fundamentally incorrect. Delegation is the process of giving an individual a specific task to complete with little direction toward the path to complete the task. Complexity is a complete system to manage a complex project by which the project manager leverages anticipated and specific human interactions in order to maximize project effectiveness.

Second, the *PMBOK® Guide* states that some material may be so general that it is not uniquely related to project management and is therefore excluded. This interpretation is possible because one can argue that complexity theory is a maturation of chaos theory, and since chaos theory is related to describing and explaining stellar movements and interactions, this material is therefore not uniquely applicable to project management (Titcomb, 1998). This argument is fundamentally flawed because there is already considerable information that shows that complexity theory is applicable to larger complex projects. Furthermore, there is already research and material that supports complexity as part of project management.

Third, the *PMBOK® Guide* states that some topics have been excluded because there is insufficient consensus on the topic. Not all project managers will agree that complexity theory is appropriate for all projects; however, it is recognized that there have been countless successful projects to date that have not utilized complexity theory or have not recognized that complexity theory was used. This may seem a reason to belittle the opinion that complexity theory is applicable to project management, and therefore complexity theory should be excluded from the *PMBOK® Guide*. But other evidence needs to be examined before accepting this opinion.

Examination of the available research into the number of successful projects that have been completed in the world will show that despite the efforts of project managers worldwide and the efforts of the *PMBOK® Guide*, there are still a large number of failed projects. In fact, the 2009 Standish's Chaos Report found that 44 percent of IT projects were challenged and 24 percent failed (Standish Group, 2009). The 2009 report saw an increase in failures from the previous Chaos report. The report also found that projects succeed for these top three reasons:

- User involvement
- Executive support
- Clear statement of requirements

While the Standish Chaos Report focused on IT project failures, this information may be extrapolated to other types of projects to indicate that over 30 percent of projects end in failure. Hence, there is still room for improvement for project managers and one such improvement would be the application of complexity theory.

Thus, the *PMBOK® Guide* recognizes generally accepted project management principles or practices; thus the goal is to better educate project managers regarding complexity theory so they can achieve greater success. All material about project management should serve to educate and illuminate individuals toward project management so that the goals of project managers worldwide are achieved.

Regardless of why, the *PMBOK® Guide* lacks any attention to complexity. When one examines all of the detailed charts, flow diagrams, and detailed descriptions in the *PMBOK® Guide*, it is clear that this is a completely linear tome. In all cases, the *PMBOK® Guide* offers solutions where A leads to B, which leads to C (Figure 2.2).

The *PMBOK® Guide* recognizes that people are an essential part of any project and people are not completely linear creatures. Processes can be established in order to better focus people toward tasks; they do not always

A → B → C

Figure 2.2 A to B yields C

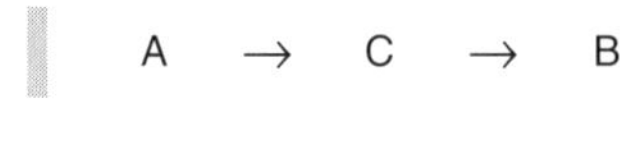

Figure 2.3 A to C yields B

operate completely in a linear universe. The *PMBOK*® *Guide* might like to believe that people are completely hierarchical and process driven, too often people are not. To this end, the *PMBOK*® *Guide* attempts to address this by trying to standardize every process to make all processes as linear as possible while recognizing that certain processes are happening at various times during a the life cycle of the project. For example, in some cases, A can lead to C, which in turn leads to B (Figure 2.3).

Initially, this example is something one would want to reject as a possibility, but one must examine this concept a little further in order to understand how it happens and why it is important. This concept exists in everyday lives; it impacts learning and is essentially a less represented and understood concept of human interaction. In order to completely understand and accept this one must explore two separate examples that are essential to understanding complexity.

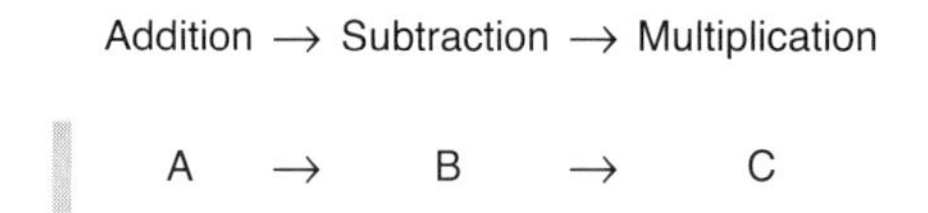

Figure 2.4 Learning addition first, subtraction second, multiplication third

In classic learning of basic math, a student is taught addition, subtraction, multiplication, and division. This is generally the accepted natural order of learning basic math. Yet there are certain cases where this is not always true (Reutzel & Cooter, 2004). Shortening this to examine just the first three elements yields the linear example in Figure 2.4.

Most people would agree that this figure 2.4 is a good representation of the natural progression of learning basic math skills for everyone. The reality is remarkably different. When one breaks down the elements of each of these math operations, one will find that addition and multiplication are closely tied together and in fact addition skills lead directly to multiplication. Individuals might intrinsically agree, because multiplication is more closely tied to addition than subtraction, most would still argue that one must learn math skills to gain a good understanding of basic math. Early learning often begins with addition and moves quickly

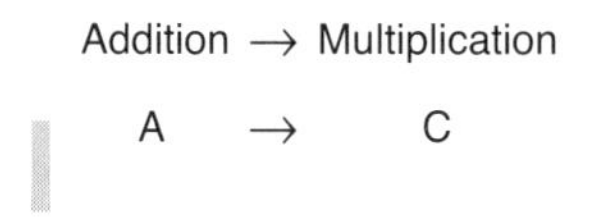

Figure 2.5 Learning addition first often leads to learning multiplication

to multiplication. This runs counter-intuitive to the order of operations; however, addition and multiplication are very tightly linked. Yet, Figure 2.5 is a clear example where A can lead to C without any intermediate step.

Again, this just seems like a potential route for some students; however, most would feel that perhaps only some gifted students could learn in this manner. Yet there is a daily reality about learning that most people chose to ignore, since it could simply be a case of memorization rather than true learning as can be also seen in some stages of reading (Reutzel & Cooter 2004). For example, shortly after a learning addition, typically the student will begin to add to together like numbers. This is in fact the basis of multiplication and is a concept that will assist them with future learning. An even more basic example is that students will typically know their 10 times table sooner than they will master subtraction. This is because once a student understands the concept of addition, the student will then practice counting by tens. This is a rather easy lesson because the pupil quickly learns that the first digit is the only change; in the process the learner quickly masters the pattern. Consider this as a real-life learning example encountered by individuals that is nonlinear and explains why complexity exists more in human interactions than most people will care to accept.

The example above is compelling but does not explain the complex nature of human interaction. It could be seen as something that perhaps exists in learning in isolated circumstances, but it does not create compelling evidence that such a concept would apply to projects. The next example of A leads to C, which leads to B will be an example that all project managers have to face and must learn to cope with regardless of how hieratical and linear their organization is.

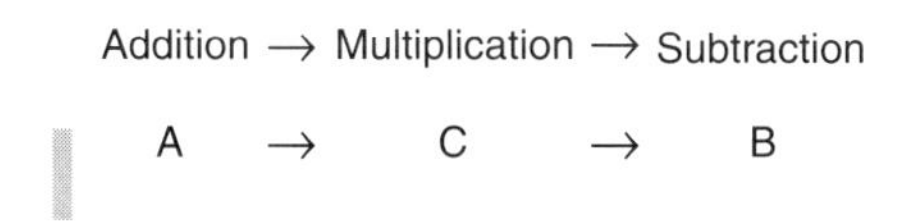

Figure 2.6 Addition leads to multiplication and then leads to subtraction

Ready → Aim → Fire

A → B → C

Figure 2.7 Ready, Aim, Fire

Figure 2.7 shows the typical progression of organizations; it is not the only possible solution. If one is looking for rational decisions based upon facts and information, consider Figure 2.7 as the correct manner to proceed. This linear type of thinking is explained in detail in the *PMBOK® Guide* (PMI, 2008) and this approach is often the correct path; however, in some cases this may not be the most direct approach. The *PMBOK® Guide*, as a standard, cannot give the appropriate consideration to external issues or complexity (this is not completely true; if you look at it there are external factors). Hence, it views these as *just another input to the project.* The guide gives no latitude to the project manager to handle unusual circumstances, such as when a project is in trouble. This is not the fault of the *PMBOK® Guide* as it is a standard that provides "good practices on most projects most of the time" (PMI, 2008, p. 3). For example, if a project is in trouble and requires intervention by a new project manager, then the typical linear approach to project management might not yield the necessary results in the required time frame. This would be considered a special case and would need to be addressed differently than the typical project.

Typically projects are in trouble for either cost reasons or time reasons. Once a project is out of control and a new project manager is assigned, there is little time for a lot of preparation. A new project manager coming in will have to make some quick and difficult decisions with limited knowledge of the project. This would then lead to acceptable nonlinear process shown in Figure 2.8.

The assumption of the incoming project manager is that some or all of the controls of systems that were in place are not functioning in a manner conducive to the success of the project and hence need to be changed in order to better drive the project toward success. As discussed

Ready → Fire → Aim

A → C → B

Figure 2.8 Ready, Fire, Aim

and supported by the Standish Chaos Report 2009, many more projects end up over budget, late, or cancelled than one would like to believe, so understanding nonlinear processes is necessary for project success. Yet, accepting this process is essential to handling smaller issues that might occur in a project. One might dismiss this example as a special case and could fall outside of the realm of the *PMBOK*® *Guide*; there certainly should be some consideration of the possibility of nonlinear thinking to solve typical problems.

These examples identify the realities of nonsequential and nonlinear thinking. When one realizes that human interaction is less mechanical and more fluid than that described in the *PMBOK*® *Guide*, one realizes the necessity of addressing complexity in the guide. The guide's lack of attention to complexity is further compounded by the lack of attention to further related topics, such as the lack of attention toward virtual projects, virtual project management communication, and virtual project leadership. The latest *PMBOK*® *Guide* (Fourth Edition) does address virtual teams but only in relation to resource management.

> Practical Tip: Consider where nonlinear thinking might help with the project. Once you find a place where it makes sense to implement, proceed with this change in order to offer people with a creative option. It does not have to be a large change to the project, but offering new ideas to the process can certainly help the organization move toward complexity. The Grand Canyon started out as a river in the desert and over time the river created the most splendid canyon in North America.

WHAT IS LACKING IN CURRENT PROJECT MANAGEMENT KNOWLEDGE REGARDING COMPLEXITY

Most would agree that the *PMBOK*® *Guide* is the standard most Western companies follow when implementing project management. There are other standards and methodologies, such as PRINCE2, adopted by the UK Government as its project management standard, and ISO 10006, a standard that monitors the quality of project management processes. None of these standards are bad but they take the safe approach and do not help to explain why projects continue to fail. Following is an analysis of the *PMBOK*® *Guide* and how it adopts complexity. It is used as an example and can be quite easily extrapolated to the other standards that use the same linear approach.

The *PMBOK*® *Guide* needs to address complexity as an essential element of any project, and by extension, the project management body of knowledge must recognize complexity as an important element of any complex project. The *PMBOK*® *Guide* needs to address certain elements of project management that are fundamental to complexity and the management of complex projects. There is material and information in the body of knowledge of project management, and there are certain elements that should be addressed in the *PMBOK*® *Guide*. In particular, four concepts that leverage complexity are either not fully explained or are treated as solely linear concepts. These four concepts are harnessing complexity to handle complex projects, leveraging complexity in communication, utilizing concepts of complexity in virtual projects, and understanding complexity as it relates to project leadership.

One of the serious omissions in the current information about project management is the understanding of complex systems that operate in a linear and nonlinear format. Traditional project management material, including the *PMBOK*® *Guide*, operates under the principle that all human systems are inherently mechanical in nature. Project management is a series of learnable processes and skills that could be applied to any new project regardless of the industry, enterprise, or society. Traditional thinking believes that all systems have predictable outputs based upon controlled inputs. This thinking is so embedded in Western thought that the project manager often forgets that there are significant uncontrollable factors in any given system. Project managers tend to ignore this material because it is considered either an aberration to the norm or simply a possibility so remote that it can be discounted as unimportant. When examining why this is such a cultural belief, one can see areas where this kind of thinking is actually a detriment. In the next section, complexity will be further explored in order to explain why a degree of chaos is essential for any complex system.

If a seasoned project manager were to explore the entire process from seed to harvest for a particular crop, that manager might describe the process as shown in Figure 2.9.

Plant seeds ⟶ Water daily for 30 days ⟶ Maintain crop health for 30 days ⟶ Harvest

Figure 2.9 Simplified farming progression

This may as a general rule describe the process at a very high level, but some external circumstances may modify this process or entirely alter the process. Anyone who has actually tried to raise a crop knows that the entire planting process is more of an art than an exact science. The physical putting of seeds into the ground has been automated and duplicated over a period of thousands of years, but there is no guarantee that the seeds will sprout. Weather, temperature, pests, and other external factors have an impact on every step of this process. The exact watering of a crop can be automated and the amount of water distributed can be regulated; this does not guarantee productivity. When one reviews the simplistic process described in Figure 2.9, one realizes that one of the important elements that is not addressed is the effect of the weather. Heavy rains, lack of sunshine, and other elements can change the growing period for the crop (Clarke, 1995). Water can be regulated when controlled; excessive waters due to rain cannot be ignored. Too much or too little sunlight can also impact the growing season for a crop. This is a fairly simplistic example yet it does offer a glimpse into complexity. The next example of a more complex crop will better illustrate that complexity exists and must be recognized as a factor in any apparently linear system.

One of the crops cultivated for thousands of years is the humble grape. On the surface, the grape appears to be a relatively simple and unremarkable product available at reasonable prices worldwide. There is still a discussion on the exact details on when this became a controlled crop; grapes have certainly been integrated into human culture at many different levels. Table grapes may be similar to the simple crop explained, but grapes can also be cultivated for a more noble purpose. Wine has been produced for thousands of years and the past hundred years have seen significant advancements in the processing of grapes into wine, yet there is still a level of art involved in this natural product (Clarke, 1995). The creation of wine is a complex project that requires not only the cooperation of nature but human intervention to guide the process. For example, there are parameters for when to harvest grapes (such as ripeness or weather conditions) but simply harvesting the grapes when they are at a certain sweetness level does not ensure quality wine. Since grapes are a natural product, it is not practical to test every bunch to determine that all bunches are at the same point of ripeness. Understanding that one is working with an imperfect system is part of the wine making process; it is the wine maker's responsibility to make the natural product into something that acceptable (and hopefully rated well by critics) in the market.

Bringing this example and back to communication in project management, one has to accept the reality that when communicating information according to prescribed and acceptable channels, the message does not always reach all recipients in the same manner and might not be accepted in the same way. This is similar to the concept that when an advertisement is circulated by mail, only a small fraction of the recipients will actually go out and purchase the product, while the majority will ignore the information. A project manager must accept that this is the reality of communication. The more that one communicates, the more likely the message will be received and understood, yet there is no certain manner to achieve 100 percent communication.

Organisms are not the same as mechanical systems. When one pushes a certain button in a properly functioning elevator, the mechanism has no latitude to go to another floor. The input (pushing the second-floor button) forces the same output (the elevator goes to the second floor). Baring a malfunction of the system, the process will be exactly the same without error. People and organizations are not so rote. Individuals often need multiple mediums of communication to deliver a message. In this regard, the *PMBOK*® *Guide* emphasizes the importance of multimedia and multivector messages and of visual cues given by individuals during a communication session. What the *PMBOK*® *Guide* does not explain is that the robust person-to-person communication is not always possible and sometimes a project manager must communicate via phone, e-mail, or text. A written report does not have the same robustness as an interactive presentation where an individual can answer specific questions in order to clarify certain points that might not have been as thoroughly explained as other points. There are ways to overcome this kind of communication limitation; the *PMBOK*® *Guide* is silent about explaining the importance of adaptive communication. The guide leaves it very open for the project manager to be creative in this complex area with only a brief caution that failed communication is the leading cause of project failure. Given the *PMBOK*® *Guide*'s recognition that this area is critically important, it would then make sense that the guide would treat this area with particular care and detail. Ultimately, the lack of details regarding communication leaves a project manager without the necessary tools to handle complex projects.

The second important area to consider, as already alluded to in the prior section, is how one must apply complexity to successful communication. The *PMBOK*® *Guide* addresses networking in the communications section (Chapter 9, 9.1.2.2; p. 222). The guide offers a great deal of excellent material that relates to project management but is considerably lacking in the area of communication. In Chapter 10 the guide stresses the importance of communication and cites communication issues as the single most important aspect of project success, but it does not offer directions with regard to complex social interaction. This social interaction is fundamental to complexity because human communication is particularly rich in subtle information. In addition, human communication and interaction are not always process based. Project communication is often not as simple as the communication outlined in the graphical representation in Chapter 10, Figure 10-1, of the *PMBOK® Guide*. The graphic implies there would be an orderly gathering of individuals and information, and from that a single font of information would spring forth and keep everyone apprised of all the matters pertaining to the project. The diagram implies that there is a linear arrangement from Section 10.1 progressing neatly through Sections 10.2, 10.3, and 10.4 ultimately to culminate in Section 10.5. There is a certain flow to the communication experience and certain details should be shared through proper channels. There is always some degree of nonofficial communication that relies upon unofficial channels.

The *PMBOK® Guide* does not take a position on this informal communication and in fact recommends that the project manager do whatever they can to suppress informal communications that affect the golden triangle of project management. A degree of static certainly exists within any communication system; a seasoned project manager will often use this to advantage. Any denizen of an organization will know that each organization has a rumor mill, which is often a clearinghouse for incorrect information, untruths, and misunderstanding by individuals with incomplete information, although the rumor mill in some organizations is often accurate and correct at forecasting future organizational events.

Most people can remember playing the childhood game of telephone where one person whispers a bit of information to the next person and so on, to find out what comes out after a dozen or so people have heard the bit of information. If this were a string of machines the message transmitted to the first machine (node) would be exactly the same message that comes out by the last machine. Anyone who has been involved in this kind of situation knows that the message that comes out on the other side is never the same as the initial message. This serves to highlight that humans are anything but 100 percent accurate when it comes to passing along information. Complexity theory recognizes this reality and hopes to educate people toward understanding this concept in order to address these problems. Understanding that organizations and people are not infallible and are in fact complex will better assist in addressing these issues in a complex project.

The third issue concerns the nature of virtual organizations and virtual teams. These are important constructs in the international and multilocation projects. Few projects are so localized that all important communication, materials, and systems are all in close proximity. Project managers have had to come to grips with this reality for quite some time. Even the most recent version of the *PMBOK® Guide* (Chapter 9, Human Resource Management section) offers minimal direction when it comes to the virtual environment. One of the shortcomings of a virtual project is that the communication mediums most commonly utilized are lacking in subtle information that is present in face-to-face communication (Duarte & Snyder, 2006).

Understanding the limitations of a project manager's communication skills is often a good first step to becoming a more effective project manager. This self-review, so important for growth, is all but ignored in the *PMBOK® Guide*. The project managers with poor communication skills are often the ones who fail although this point is identified in the *PMBOK® Guide*, the unfortunate truth is that no further details are given to overcome this shortcoming. It is an unfortunate situation when a problem is recognized but no solution is given. Project managers turn to the *PMBOK® Guide* for answers, not to find likely causes of failure with no alternatives or solutions.

The *PMBOK® Guide* offers an excellent overview of the fundamentals of the process of communication. Chapter 10 of the guide puts more emphasis upon the avenues of communication—such as to whom and where to distribute information—than it does on educating the project manager upon the manner, style, and types of communication. The guide offers excellent advice regarding the importance of reports and the timeliness of this type of communication. However, it does not offer any direction about handling situations where communication is already a problem and how to correct these types of problems. Appendix G of the guide identifies communication as one of the most important factors in project success but offers little in the way of practical advice about how this communication should be implemented. Some excellent graphics and charts in the guide offer advice on when and where to communicate, as well as indicating the importance of a multidimensional approach to communication (in person, reports, e-mails, and the like). However, it does not offer a clear blueprint on what should be done during a project. It is almost as if the project manager should already know these techniques from past experience and the *PMBOK® Guide* is only trying to offer an additional framework, rather than presenting a solid foundation for the project manager to work from.

> Practical Tip: Review Chapter 10 of the *PMBOK® Guide* and reflect upon the importance of communication. Consider how this information could apply to your current project or future project. Also consider what other resources about communication are necessary and work toward gathering those resources as soon as possible.

The project management standards should include more information about the virtual environment since it does not even address this important aspect of project management. Organizations have grown and budgets are shrinking. A project manager cannot continually travel from location to location to monitor a project. No longer are projects isolated in a single geographical area. It is not uncommon to have a project's final deliverable in more than one location, to have procurements for the project coming from other countries, and to have project team members in different countries all working toward the same goal. The global village is a reality for project managers, and those who are capable of handling, monitoring, and communicating with a team that is not physically located with them will become more in demand. The *PMBOK® Guide* needs to include information of this nature in order to remain a current standard. Perhaps virtual projects were not as common in the past, but the reality in today's environment is that virtual projects are the silent majority.

The *PMBOK® Guide* presents some discussion about how communication might look in the virtual environment, yet it offers minimal guidance regarding how to handle these kinds of environmental challenges, such as how a project manager can successfully manage a project located in multiple locations. Addressing virtual projects and projects whose team members are spread around the globe is an important matter that needs to be addressed sooner rather than later.

Project leadership is an important aspect of project management, and for discussion in this section they should be considered the same. What should not be considered the same is the difference between project leadership and virtual project leadership. Applying complexity to virtual project leadership is yet another consideration that needs to be examined. Given this starting point, the *PMBOK*® *Guide* offers excellent documentation, information, and recommendations for project leadership utilizing a contingency-based leadership style. The material in the *PMBOK*® *Guide* is very applicable with regard to managing a traditional project where the project manager is on site and the team is all located in the same general area or location. This information is not only accurate but it has been researched and supported over many different organizations and research cases. The dispute is not with this material but with the lack of the same treatment of virtual project leadership.

Virtual project leadership is different than leadership in a traditional environment (Duarte & Snyder, 2006). There are some similarities, yet the *PMBOK*® *Guide* does not account for these differences. These differences are more than what is addressed in the *PMBOK*® *Guide*. The guide blankets this type of information under political and cultural awareness in Appendix G.7. The guide dedicates an entire paragraph to cautioning project managers about the global environment and the environment of cultural diversity. The only recommendation to end this is that the project manager should know the various team members and use good communication planning as part of the overall project. Good communication planning is addressed elsewhere in the *PMBOK*® *Guide*, which does not offer any practical recommendations about handling the leadership of a culturally diverse or a geographically dispersed team. Having an understanding of the differences is essential to a successful project and complexity is an excellent vehicle to facilitate this success.

To start, one must expand upon the recommendation found in the *PMBOK*® *Guide*. The advice given in Appendix G7 of the guide is actually a good point of consideration; nevertheless, it should only be taken as a first step in the process. There is certainly a need to create a good initial communication plan. The problem is that this advice is simply not enough. The reality is that given any complex project, any initial plan that is developed will usually change many times throughout the project. A virtual project is even more dynamic than a traditional project so one will need to remain flexible in order to be successful. Furthermore, in a virtual project the lines of communication are often more tangled and less affected by hierarchy due to the apparent chaotic nature of communication. One must expect that there will be more static in the communication of a virtual organization than in a traditional one. Communication will lack the robust nature of face-to-face communication and often clarity is slow to achieve. Since virtual teams might be in widely different time zones, it can make communication slow (Duarte & Snyder, 2006). This slowed communication can appear as static as different locations vie to understand the messages being issued from above. Because team members will receive communication in their own time zones, these delays will create temporary voids of knowledge that can impact a project. This will create potentially negative information to be filled into these voids. If a leader is not careful, the organization will assume the worst and communication will become a bottleneck as individuals become fearful. This is where some knowledge of complexity can become instrumental in success.

A virtual leader must accept that there will be a level of disorder or chaos in the lines of communication. Leader can do their best to plan for the clearest lines of communication and have multiple contingency plans so that everyone is fully up to date at all times. A good project manager will make sure that communication is sent out at the same time to different groups and the good virtual project manager will be meticulous about following up with key members to make sure that critical communication is understood in a timely fashion.

Consider an astronomy metaphor to help guide you in understanding how a project can be successful and still utilize complexity. Our solar system is composed of eight planets (including the earth) that revolve around the sun. Beyond the eighth planet are several dwarf planets and then a meteor belt. This orderly view of our solar system is good for a general understanding of our local neighbors; it does not completely describe the situation. The reality is that a considerable amount of activity occurs between these planets. Anyone who has looked at the craters on the moon or seen a falling star understands that the solar system contains a lot of material in addition to the eight planets. In fact, as discussed before, complexity theory was originally applied to celestial bodies in hopes of describing and understanding some of the movements of these celestial bodies (Byrne, 1998).

Now, taking this back to virtual leadership, a leader can certainly help keep eight individuals orbiting in the right direction and moving along a set path, but unanticipated or unpredictable static will always appear. A virtual leader understands this concept and not only plans for it, but continually updates the project to accommodate this type of communication. In some cases, it is best to suppress this type of static (if negative, for example), but in other cases it can be used to help support the project. A virtual leader must not only understand that such interference exists but must understand that such interference is natural. The goal is not always to suppress this type of behavior but to shape it to a productive force in the project. The more one can leverage all of the forces at the disposal of the project manager, the better the project will become.

Additionally, there is evidence that virtual leadership must be transformational in order to be successful, and transformational is an essential element of complexity. Unlike a traditional project where everyone is located in one area and is relatively proximate, a virtual project does not have this kind of direct social connectivity. A virtual leader must understand that creating something new from something old will require a degree of transformation. The *PMBOK*® *Guide* often cites linear progression toward change while transformational change is often more successful. Often a radical change is necessary for the success of a project. Individuals are often mystified by how they will go from one state to another. This is actually quite a common mystery in the physical sciences.

Radical transformation is no different than the change of the states of matter. All matter can exist as a solid, liquid, or gas, depending upon the temperature or pressure exerted upon it. The most common example of this transformation of matter is seen in water. When liquid water is brought to a temperature of 32 degrees, it transforms into a solid, ice. Ice remains a solid, but what is interesting is that ice and water can exist in the same area. Anyone who knows about the tragedy of *HMS Titanic* understands that water and ice can coexist. There can be huge icebergs of solid water along with vast quantities of liquid water in the same area—the ice and the water will both exist at almost exactly the same temperature. Anyone who has seen a lake freeze knows that most of the time there will be a thick layer of ice on top, while the lower portion of the lake remains as liquid.

At the other extreme, anyone who has boiled water to make pasta has seen liquid water transform into a gas. Again, water exists in a state where the liquid is hot enough to become a gas but some remains as a liquid. A project manager must learn that a team will transform in precisely the same manner and must do their best to move this process along. When a virtual leader moves a team to a new plateau, some team members will embrace the change and move to the different state, while others will remain at the old state. The challenge becomes that the virtual leader will have to bring everyone to that new state. It is not an easy task as some people will resist and others will find plausible reasons to remain static, but in the end it is the leader who brings about the organizational change that is necessary.

WHAT THE *PMBOK®* *GUIDE* SHOULD INCLUDE ABOUT COMPLEXITY

Complexity is a current topic that is growing in importance. Globally there is a wave of reawakening that complex systems exist and cannot be predicted. Just as certain experts are studying the global temperature and trying to make sense of this in relation to the complex global system, project managers are beginning to realize that certain social environmental issues cannot be predicted by conventional linear thinking. Humans have been trying to control or make sense of their environment since the first human tribe came together for mutual protection. This concept of understanding is something that is deeply rooted in our psyche. Just as primitive man created codes, mythologies, and laws in order to improve order, humans have continued in this direction to attempt to explain all of the mysteries of the universe. In fact, linear thinking is becoming more of a hindrance in today's world than ever before. Urbanization has taken most individuals out of a naturally connected world where it was apparent that certain systems could not be predicted or controlled. Urbanization has created the veneer of control over the environment. This control became more of a belief once mankind had achieved a level of technology that would allow for the systematic destruction not only of mankind but the technology to practically destroy all life on the planet. This kind of control has created a belief that there is always a linear explanation for all problems in the universe.

Considering the reality that the *PMBOK®* *Guide* has done an excellent job at giving structure to project management over the years, it becomes difficult to incorporate sections of nonlinear thinking. This would seem almost counterproductive to the attempt by the Project Management Institute™ (PMI) and the *PMBOK®* *Guide* to create greater structure, science, and sense to the field of project management. The *PMBOK®* *Guide* should address "good practices on most projects most of the times" (PMI, 2008); the *PMBOK®* *Guide* and other project management standards and methodologies should dedicate a full section toward the concept of complexity and how it fits into project management. This section of the guide should also include a more modern and successful leadership approach for project managers.

In order to make this section on complexity effective, the *PMBOK®* would need to dedicate an introductory section to explain the history of complexity and chaos theory in order to introduce the concept to project managers. Since the existing school of thought is that all project management must be linear and orderly, this should explain how chaos is a naturally occurring feature in the universe and a degree of chaos exists in all complex systems. This requires some explanations from researched projects that applied complexity in a successful manner. The most obvious application is to any type of human organization. By definition, human organizations can be structured, and processes and procedures can be codified but behavior cannot be fully controlled by any process, system, or person. Humans operate as individuals and they have their own thoughts, ideas, interpretations, and agendas; this cannot be overlooked. Addressing these aspects can create a more powerful structure that can better harness an organization's success.

Chaos theory and complexity theory have been evolving in recent years, so it is critically important that the *PMBOK*® *Guide* establish a common nomenclature for complexity as it applies to project management. The guide requires some further clarity to avoid using complexity as a reference to a complex project. A complex project should be kept separate from complexity, and the guide should avoid phrases such as "Projects vary in size and complexity" (PMI, 2008, p. 16) because this statement is about a complex project rather than about the application of complexity to a project. This clarity of nomenclature about complexity theory is essential to project managers because it helps create a consistent vocabulary that can then be applied to successful projects. Because most project managers do not currently understand complexity, it is important to make the concept clear to everyone to avoid confusion. The importance of this element should not be overlooked, because the creation of a powerful, clear, and concise vocabulary will allow for projects to reach higher levels of achievement than in the past.

The next section should address complexity and project management. This would require addressing the lack of treatment of this area and how project management has evolved to embrace this type of thinking. This would not be a clear case of the white hats and the black hats coming together to agree that everyone should wear gray hats to respect the history and tradition of both groups. This would be more of an acceptance of an entirely different school of thought that can be successfully integrated into the mainstream of project management thinking. Historically, the *PMBOK*® *Guide* has been put together by committee and these are made up of practitioners that do not do formal research, there certainly is room for creative thought. This certainly would take time for acceptance, yet, a document such as the *PMBOK® Guide* should address all relevant and important thinking when pertaining to project management. The failure to address existing research and ideas can be viewed as an omission and can certainly detract from the success of the *PMBOK® Guide*.

The *PMBOK® Guide* needs to expand its thinking in order to embrace other successful schools of thought in order to remain current. For example, the *PMBOK® Guide* has several new editions in the past several years and this is because of the needs of the *PMBOK® Guide* to remain current with the new rising tides within project management. In fact, the *PMBOK® Guide* has had four editions in just twelve years. This is clearly an indication that the needs of the project management community is not only growing but is evolving. The *PMBOK® Guide* boasts an extensive number of contributors, experts who span multiple areas and disciplines with both formal and experience-based backgrounds. Appendix G, which is the only area that addresses interpersonal skills, cites only six references, two which are eighteen years out of date, two of which are thirteen years out of date, one that is eleven years out of date, and one that could be considered as current at three years old. This makes the average age of this research to be a little over twelve years old. Since there have been four editions in twelve years, it almost appears that Appendix G used material that was part of the first edition of the *PMBOK® Guide* and never attempted to update this section.

For the project management community to remain current, new ideas must be injected into the community to keep the ideas applicable and fresh to new projects. Furthermore, the *PMBOK® Guide* must retain its position as an eminent standard of project management. The Project Management Institute™ (PMI) is clearly the global leader in project management; it should not consider that this position will be kept if it does not continue to evolve to address the changing needs of its constituency.

A further section should include material that brings the theoretical nature of complexity theory to the forefront of project management. It must show how a project can be successful by embracing the ideas and concepts of complexity. Complexity needs to have a direct connection to project management. Project management must thrive by learning to apply the linear thinking presented in the current *PMBOK® Guide* and also to apply nonlinear thinking of complexity. Assisting project managers to utilize both concepts will assist them in understanding how complex systems operate. As discussed before, process management is all about linear thinking and linear decisions; individuals or groups often do not behave in this manner.

Complexity is about anticipating the unexpected in projects. When a plane goes from Los Angeles to Miami it leaves with a set and registered flight plan, but will make numerous course corrections and adjustments before it arrives at its final destination. A plane departs with a clear plan set for the flight but will rarely, if ever, remain on the initial course the entire duration of the trip. The path may seem clear and linear but when one examines the exact process there is a degree of complexity to the flight. The passengers on board will have scant understanding that they have been off course for part of the flight, but the diligent flight crew will understand the need for these kinds of changes. It is this exact concept that must be reflected in project management. Adhering rigidly to a set plan can often stifle or delay a project. In some ways, a linear project manager may just see this as having a flexible plan or may consider these changes as contingencies; the actual reality is slightly different.

Complexity in project management is about applying the concept that linear plans are subject to human and environmental adjustments, just as a flight crew adjusts for weather patterns that may disrupt the flight but nevertheless all the passengers still arrive at the final destination. Applying this kind of thinking is harder than it would appear because it creates some degree of flexibility that might not be acceptable to a traditional project manager. Often a project is successful because of its rigidity and the failure of the project is often blamed on the lack of a firm policy, procedure, or process. Project managers called upon to repair failing projects almost always clamp down on all policies, procedures, and processes. Many times, this kind of openness creates disruption within a project. This may be true in some cases, but it certainly destroys any type of creativity or creative thought, which almost always goes by the wayside on a failing project. However, complexity thinking may be the correct time to put it to use.

Some project managers relegate complexity to nothing more than a dressed-up version of empowerment or delegation. After all, the example of the flight from Los Angeles to Miami could be interpreted in a manner that avoids the application of complexity. A seasoned project manager would take the earlier example and dismiss complexity as just giving the pilot some latitude in interpreting orders (the flight plan, or if you prefer, the project plan). The flight crew has the responsibility to bring a planeload of 150 passengers from Los Angeles to Miami. The flight crew is empowered, within a limited set of circumstances, to alter their flight plan. They have control of the voyage in order to make decisions of safety, security, or timing. Hence, the flight crew is empowered to make a narrow set of adjustments as long as they adhere to the vision of arriving in Miami at a predetermined time. This example appears similar on the surface to the earlier one about complexity, but this is where the comparison ends.

Complexity is about the degree of chaos inherent in human interaction. It is a level of acceptance that chaos is continually occurring and rejuvenating. The examples about the flight from Los Angeles to Miami actually discuss and review two totally unrelated elements, yet they appear to be identical concepts. Intellectually, people seek to create order and connections to unlike concepts so that they can be cataloged and parsed later in a manner that is helpful to future understanding. So, one might ask, how are these examples different when they seem so tantalizingly similar?

The first example about complexity was a narrow concept solely related to the idea of course corrections. The second example distorted that concept in a manner to make it more approachable. The second example about empowerment is entirely correct; the concept of course corrections becomes a minor detail in the entire process. In fact, the whole idea of course corrections becomes the will and right of the pilot rather than an environmental necessity. The concept of empowerment is about allowing people some level of interpretation about orders and holding them accountable for actions. Complexity is about the manner in which human thought and human-controlled systems have a degree of unpredictability due to the fact that everyone is different. The recent "Miracle on the Hudson" is an example of how one heroic pilot averted disaster and saved the lives of hundreds of people. A different pilot might have had a less successful landing under such difficult conditions.

Different experiences create different people. In fact, identical twins, who from a scientific standpoint are basically clones, are actually different and continue to develop differently. Identical twins at birth are different weights and will continue to develop at different levels. Identical twins can even develop hereditary disorders at different rates, so there is a degree of unpredictability in people who are considered genetically identical. This point serves to illustrate that people do not operate the same as machines, nor do they all behave in a manner that is totally predictable. Complexity theory is a way of making use of this type of information in a manner that will assist in the future. Just as the course corrections analogy will assist us in understanding that there is a degree of chaos in apparently orderly processes, the goal is to apply this kind of thinking in a way that allows the project manager to strengthen the project with this kind of knowledge.

The challenge then becomes how to achieve this goal. Several books have tried to explain how complexity can assist, but there are scant recommendations for project managers. Some complexity scholars find that the social sciences should avoid using statistics to predict human behavior. Sometimes examining a single element of society as a barometer for society makes for a compelling argument, but the problem remains: how does one apply complexity to projects?

The most important elements of complexity that apply to project management are transformational and dynamic. These are the two elements that should be addressed within the body of the *PMBOK® Guide*. First, a project leader and the project must be transformational in manifestation. Second, the project plan, the project leader, and the project team must be dynamic. Creating an understanding of these concepts and bringing them to project managers would improve the body of knowledge for all project managers. This would further develop projects and project managers that are better equipped to deal with the extremely dynamic nature of complex projects and project management.

This concept of transformational leadership is unfortunately in current conflict with the leadership proposed in the current *PMBOK® Guide*. The current guide does not articulate the current body of knowledge with regard to leadership because it considers the topic too vast to cover in depth (p. 240). The *PMBOK® Guide*'s recommendations about leadership are based on the contingency theory of leadership.

The contingency theory of leadership centers on the belief that management is not an overseer whose primary job is to ensure that labor remains focused on their tasks, but instead on the idea that management and labor must learn to communicate and work together effectively in order to maintain order and peak productivity (Bass, 1990). This concept of group theory leads to the development of the contingency theory of leadership. Contingency offers leadership as a function of task-oriented and relationship-oriented situations, which relate to group performance. Fred Fiedler's (b.1922) contingency theory is the basis for much of these ideas presented in the *PMBOK® Guide*. This is the kind of leadership where senior officers expect junior officers to submit reports, consult superiors (stakeholders) for certain decisions, and handle scheduling matters (planning).

Although contingency leadership is well documented and supported by years of research, it is not always the most effective manner of management (Bass, 1990). Contingency theory does not offer any direction about how interpersonal skills might make for a more effective leader. As this recommendation is to update the current knowledge, rather than just the inclusion of material on complexity, this update requires some additional support in order require this change.

This is exactly the kind of leadership described in the *PMBOK® Guide* in Appendix G, which offers more information on the interpersonal skills required for project managers. This type of leadership is certainly better than some of the others discussed earlier, but has not been shown as successful as transformational leaders regarding project management.

Again, the *PMBOK® Guide* still agrees that there is a lot of additional information available; the guide should offer the best and most current information available. Since several studies have supported the idea that transformational leaders are more successful project managers, it would be in the best interest of the *PMBOK® Guide* to make this kind of modification. At the very least, the guide should offer transformational leadership as an alternative to contingency-type leadership. There is agreement that a leadership style is not the only reason for project success; there is certainly evidence that certain leadership styles are more effective. Clearly, many successful projects have had leaders who used leadership styles other than contingency. There is evidence that supports a greater success rate of transformational leadership; the *PMBOK® Guide* should consider this update to better serve their constituency.

> Practical Tip: Reflect upon your leadership style and consider if you operate from a theory of contingency leadership. If you do operate from contingency leadership theory, then consider other leadership styles to better understand what other leadership styles and ideas are available. Keep in mind that you can merge, meld, and otherwise combine leadership styles to create the one that best suits your needs.

COMPLEXITY BEYOND WHAT THE *PMBOK® GUIDE* SHOULD INCLUDE

The complex and multidimensional nature of virtual projects is forcing project managers to utilize a new standard of rules where the old body of knowledge is no longer adequate. In order to address this challenge, projects must embrace a degree of complexity theory in order to better organize. In some cases, culture is forcing complexity theory to become the solution of choice because it is better able to focus a virtual project. It is precisely these types of situations that are not addressed by the *PMBOK® Guide*. As businesses continue to contract and management is required to address larger and larger scope projects and larger and larger teams, the *PMBOK® Guide* must also evolve to address these pressing changes.

The 2009 Standish Chaos report confirmed anecdotally that projects were not providing results, either financially or in the timeframe promised. This has also been confirmed by academic research, where the ranges are from 20 percent for traditional software projects to as high as 80 percent for virtual projects, depending on the definition of the failed project. Therefore, it is time for academia and the practitioner to review alternatives to advance the practice of project management. Goldratt's critical-chain methodology has not gained traction in the world of project management, although it had some valuable assets. The Project Management Institute™ (PMI) has acknowledged this methodology in the *PMBOK® Guide* (PMI, 2008) as an acceptable method to measure progress. The major scheduling software tools do not allow for this type of scheduling. This has created a serious disconnect between the available tools and the *PMBOK® Guide*.

The field of project management is maturing and it needs to find alternate theories that may assist the project manager in delivering within the golden triangle of budget, scope/quality, and schedule. The *PMBOK® Guide* does offer some guidance toward outside tools and techniques with regard to leadership and social networking, but there is little direction in this area. Others would oppose the *PMBOK® Guide* recommending certain leadership strategies, let alone recognizing the importance of social networking systems; the decision to remain silent in these areas is one that has vast implications. Not only does it leave the project manager adrift in the sea of leadership styles available to choose from, but it also leaves the door open for a project manager to justify choosing a leadership style that is out of date or clearly a poor choice for a particular project.

The virtual environment creates microcosms of culture (Cooke-Davies, Cicmil, Crawford, & Richardson, 2007) within the virtual environment similar to the business social networks, LinkedIn™ or Plaxo™ or to the more social networks such as MySpace™ or Facebook™. This may be due to individuals gravitating together due to like experiences and backgrounds. This is somewhat hampered in a virtual environment because names, locations, company, team, and other interests can highlight the similarities. The project manager needs to promote these social networks without forcing them. Often the project manager can promote these smaller subgroups by assigning smaller teams to tasks. A project manager should assign two people to work together on a specific deliverable. The deliverable can be broken down into elements and each person can be responsible for an element, but the entire deliverable becomes the goal of the small team. This will help individuals to pool their resources to meet the deadline.

Academic research (Cooke-Davies, Cicmil, Crawford, & Richardson, 2007) and anecdotal evidence (Standish Report, 2009) have demonstrated that project management processes appear not to be enough to maintain the sanctity of the project golden triangle, cost, scope/quality, and schedule, especially with complex projects. Even seasoned project managers agree that all parts of a project cannot and should not be tightly controlled. A virtual project is similar to the beehive or ant colony—a microcosm of the organization. There are many things going on around the queen and things are being done routinely, abiding by processes. A level of cultural contingency must take place in order to ensure that tasks do not halt while the team waits for clarification. Just as social networking sites have restrictions and limitations regarding access and exposure, these same kinds of restrictions and limitations must be enforced upon a virtual community that is focused upon a project.

Beyond leadership and social networking, which are part of complexity, two other areas must be addressed in greater detail. The *PMBOK® Guide* only tangentially addresses the concepts of program management and portfolio management. The *PMBOK® Guide* offers additional tools in the areas of program and portfolio management; there is actually no direction toward these documents in the *PMBOK® Guide*. These concepts are addressed as higher layers of project management, but then the guide fails to offer any guidance about these additional tools. It is interesting to note that these meta layers are actually areas that can be better managed with complexity, but the *PMBOK® Guide* offers scant information in these areas. The introductory section of the guide defines and even diagrams these other layers of project management, but these are not clearly addressed. These could be thought of as the more executive layers of project management; yet the reality is that an understanding of all layers of management is required. In fact, the *PMBOK® Guide* addresses the notion of stakeholders in a concise manner yet feels that those other layers of project management are not necessary.

> Practical Tip: Join a virtual social network to expand your network. There are many to choose from but all have the potential to connect you with other colleagues, with people with similar interests, and most of all with future leads and contacts. There is something to the old adage that it is not what you know, but who you know.

HOW OTHER PROJECT MANAGEMENT ASSOCIATIONS ADDRESS COMPLEXITY

Complexity remains an elusive concept to project management globally. There is growing acceptance and research regarding complexity but there is still no universal embrace of the concept and application to project management. No formal project management organization has adopted complexity as part of their standards, yet many research-based organizations are recognizing the importance of this field. This recognition is certainly making the field interesting to business; it lacks any kind of formal recognition in academia. There are a growing number of project management certificate programs through academic institutions, yet none of these academic institutions are offering any formal curriculum in complexity. This makes for a difficult quandary because there is growing research internationally in the field of complexity and there is growing interest by business to embrace complexity to stem the tide of growing economic uncertainty, yet formal academic institutions are unwilling to embrace this field as it relates to business.

Certainly part of the reason that complexity has not become part of the project management curriculum of colleges and universities is the lack of interest in this area by PMI. PMI has not formally acknowledged complexity as a viable area of project management study. Academic organizations are forced into a consumer-driven role with project management. As many institutions have only recently embraced project management as an academic field, it has become important for these institutions to follow the lead of PMI in order to achieve academic accreditation.

As previously mentioned, academic institutions are consumer-based in relation to project management and if the consumer is looking to achieve a Project Management Professional (PMP) status through Project Management Institute™ (PMI), the institution must then offer a program that supports the student in this manner. Thus, one will find that many of these project management degree and certificate programs are designed to assist the learner in eventually achieving PMP accreditation. Since the PMP test is designed around the current *PMBOK*®, the curriculum of these institutions will then focus on the material from the *PMBOK*®. Currently, the *PMBOK*® does not make any reference to complexity theory or how complexity is part of project management; sadly, one will not find complexity as part of the curriculum from any college or university that is offering a project management degree or certificate. The failure of most of these organizations is the lack of sound research beyond the *PMBOK® Guide* regarding to project management.

Had these research-based institutions examined the current body of knowledge of project management, they would certainly find current references to complexity and understand how this knowledge fits within current business practices. Understanding this connection is important in order to offer a well-rounded and research-based curriculum rather than just creating a curriculum that will assist a budding project manager complete the PMP test requirements. The research-based management and project management organizations outside of PMI offer significant supporting information that complexity is an important aspect of project management and virtual project management.

What becomes the most interesting point about this omission is that many of the same academic institutions that offer project management certificates or degrees that do not include material about complexity also support project management research into complexity. Two such organizations that are clearly supporting complexity theory and project management are the Academy of Management™ (AOM) and the Society for the Advancement of Management™ (SAM). Both of these organizations are directly supported by some of the top universities in business and management, yet research in complexity done by these academic and practitioner-based societies fail to impact the curriculums of the academic institutions that support them. This might be explained away as complexity being new and untested in project management; one actually only finds this view in academia within the United States. There is much greater acceptance of complexity as part of management and project management and in particular virtual project management within business. Once one starts to review how complexity is viewed outside of the United States, one finds that there is a much greater acceptance of complexity not only by business but by academic and public institutions.

Practical Tip: Consider joining a management or project management organization. Joining a group like Project Management Institute™ (PMI), Society for the Advancement of Management™ (SAM), or the Academy of Management™ (AOM) can improve your network and potentially expand your resources for projects and teams.

INTERNATIONAL RESEARCH NETWORK ON ORGANIZING BY PROJECTS

The International Research Network on Organizing by Projects™ (IRNOP) has been supporting research for years regarding complexity and this organization continues to contribute to the development of the emerging body of knowledge regarding complexity in project management. IRNOP—which is supported by the Umea School of Business at Umea University in Sweden, and IRNOP conferences are organized by the Berlin Institute of Technology—offers a research platform supported by European universities. IRNOP conferences are probably the better known international organization that is currently supporting research in complexity and project management because it has two major sponsors: Gesellschaft für Projektmangement (GSM), which is the German Association for project managers, and PMI. It is interesting that PMI has become a sponsor of the IRNOP conferences, because these conferences often offer alternative research in project management. PMI does not support complexity research through the *PMBOK*®. It is important to note that PMI recognizes this type of research as important to the global project management community.

IRNOP has continued to support leading-edge research in the field of project management. IRNOP has been supporting research for years that has offered a clear alternative to the existing body of knowledge of project management. By supporting alternative project management research, this organization has offered a forum to discuss various important matters that relate to complexity and project management. The 2009 IRNOP conference, held in Berlin, specifically called for research papers related to complexity and risk management in projects. Specifically, IRNOP was addressing need for research in managing complexity and risk in projects. This is a clear indication that complexity in project management is a fundamentally sound concept that requires further attention by the project management community. This call for papers about complexity is designed to offer more research in this area to better inform their constituency about complexity and how it relates to project management and specifically to risk.

INTERNATIONAL PROJECT MANAGEMENT ASSOCIATION

Engineering and Physical Sciences Research Council (EPSRC) is a composition of seven public research councils in the UK. This strategic partnership invests annually about 2.8 billion UK pounds (approximately US$4.4 billion) toward research in various areas of importance as guided by the governing council. It is important to note that these funds are how EPSRC derives its research and operating budget from public funds from the UK government. Because of this public nature and the large expenditure of funds by the government, EPSRC dedicates its research to information that will increase the body and the wealth of knowledge with regard to engineering and the physical sciences, but this funding is necessary for increasing and demonstrating economic impact. Both of these organizations have been supporting complexity research for years and both have contributed to the development of the emerging body of knowledge regarding complexity in project management.

EPSRC is another organization that offers alternative views to project management. EPSRC operates from a sound belief that *theory leads to practice and experience generates knowledge*. This fundamental belief offers a platform that is lacking in other project management organizations. Instead of limiting an organization to practitioner-based knowledge and experience, EPSRC embeds theory and practice together and connects it to practitioner-based knowledge. This is of particular interest because this is a different set of beliefs held by many other project management institutions. Since EPSRC is based upon the physical sciences and engineering, and is further influenced by the cross-disciplinary nature of the seven councils, it creates a more fluid and dynamic organization that is looking for solutions to assist with developing the future.

From 2004 to 2006, EPSRC funded a research project, Rethink Project Management. This project was designed to update the underlying theory of project management to encompass more current and conventional project management theory. This research by EPSRC was a research network of academics and practitioners who would be able to better identify the new direction for project management in the twenty-first century. The high-level goals of this research project were to move project management away from these older management ideas, to move away from the narrowly based project management theory that has become mainstream, to move project managers from Gantt Chart management, and to address the growing criticism of different project management bodies of knowledge, such as the *PMBOK® Guide*.

There were several interesting findings from this research; the most compelling point was Direction One of the research. Direction One states, "The Lifecycle Model of Projects and Project management must move towards Theories of the Complexity of Projects and Project management" (Winter & Smith, 2006, p. 5). This first direction clearly shows that the future of project management lies with complexity and no longer with just the lifecycle model. Despite the revolutionary notion of moving toward complexity, the EPSCR researchers are clear to point out that this movement is an enhancement and not a replacement. The research points to project managers embracing concepts of complexity in order to better handle the projects of the future. The conclusion of Direction One is that there is a need for multiple images of project management in order to guide actions of management, regardless of the level. Project managers need to refer to the lifecycle model (as expounded by the *PMBOK® Guide*), but they also need to understand other theories and practice in order to handle more complex and virtual projects.

The other compelling finding from this research is that project management practitioners must move from being trained technicians toward becoming reflective practitioners (Winter & Smith, 2006, p. 5). This is an interesting finding because there is an undercurrent in project management that its institutions should direct the community. Some consider this necessary for the community to grow beyond the confines of the institution; the community must begin to direct the institutions. What is interesting about this is that this process mirrors the cultural evolution of many nations and states. This even mirrors the history of leadership theory by the evolution of leadership thinking going from a top-down hierarchical manner of management to one that is bottom up, dynamic, and based more upon a network of confederation of members. There is no doubt that for project management to evolve and develop that it will require more of a grass-roots type of movement within project management. This kind of movement can only be done when enough practitioners are finding the current deficiencies in the existing institutional body of knowledge and require that the institution update their material to include other vital material that addresses the needs of the constituency. It may take PMI some time to embrace these findings, but there is no doubt that eventually there will be a need to address complexity in greater detail in the future.

> Practical Tip: Consider that there is already acceptance of complexity theory internationally. If there is not more acceptance of complexity in the United States, then project management expertise might fall behind that of other nations.

PRACTICAL APPLICATION IS MISSING AND WHERE CAN IT BE FOUND?

Many project managers, especially those on large, complicated projects, would state that the *PMBOK® Guide* standards are not enough to manage a project let alone a virtual project. Competent, successful project managers on these large projects would also suggest that the project manager cannot and should not control all aspects of the project. A certain amount of chaos or complexity should be allowed to happen on large projects. As projects continue to grow in size, companies continue to outsource, the world continues to see people cross borders, and the project manager adds the additional complexity of culture. The project manager is now faced with massive lines of communications because of the sheer number of individuals on the team, but the project manager may be dealing with language barriers, ethical issues, technology dilemmas, and cultural tensions, among other issues that happen between people.

There is sufficient evidence to suggest that a paradigm shift is needed for managing complex projects, which encompasses virtual projects, from the traditional project management processes (Cooke-Davies, Cicmil, Crawford, & Richardson, 2007; Singh, & Singh, 2002; Leban, 2003). PMI does not suggest a formula for managing complexity. In the past few years, there has been an upward growth of projects as organizations continue to grow in size, companies continue to outsource, the world continues to see people cross borders, and the project manager must deal with the additional complexity of culture. The project manager is now faced with significant lines of communication because of the sheer number of individuals on the team. Consider that a project manager may be dealing with language barriers, ethical issues, technology dilemmas, and cultural tensions, among other issues that happen between people (Krajewski & Ritzman, 1996).

Project managers must allow individuals to resolve their own problems in order to stimulate the development of autonomous work teams. Not all situations are black and white, especially with the numerous cultures involved, technologies involved, and the situation that has to be resolved. Some situations may need to be settled by several telephone calls rather than having to create a new process/procedure to address an unusual circumstance. Some cases the project manager does need to review, such as when laws and regulations are a factor. In those circumstances, the project manager should insist upon involvement, and then the creation of a process and procedure would be a requirement.

Some situations can be dealt with through processes and procedures but research has shown that more likely they should not be. Sometimes even the project manager cannot account for all situations, and the project manager cannot be everywhere at all times, particularly in a virtual project that has project teams in many different countries and in many different time zones. Hence, there is clearly a need to locate sources of information to support project managers that have an interest in becoming more successful with complex and difficult projects.

Most project managers have been educated to search for these functional and research-based tools through PMI. Project managers will be hard pressed to find much in the way of support and documentation regarding complexity. Plenty of information and tools are available through PMI, but project managers may find themselves without any tools to support a difficult and complex virtual project. PMI may find that that offering more support of complexity will better serve their constituency; for now, the seasoned project manager will have to resort to searching the files and articles of research organizations, international organizations, and other project management publications.

Practical Tip: Consider finding other resources regarding complexity in other industries because this kind of knowledge can often be applied with great success to other areas. It is not an exact science of applying this kind of knowledge across the board, but it does offer some new ideas for a project.

CHAPTER SUMMARY

Although the *PMBOK®* *Guide* is lacking in the area of complexity theory, there is still an opportunity to include this kind of information in a future revision of the guide. Clearly many leadership techniques are available to project managers and the more that are explored in the *PMBOK®* *Guide*, the greater number of resources that will be available to future project managers. There is a wealth of resources for project managers out there in the form of different project management organizations in all parts of the world. A project manager has to determine if they are interested in material for virtual or nonvirtual organizations and that should narrow down the search enough to find an appropriate organization. If a project manager is unsure of what or where to begin his or her search, the first place would be a review of the Project Management Institute™ (PMI). Regardless of the resources that the project manager might need, PMI can offer an appropriate resource or at least direct the project manager to an alternative source of information.

REFERENCES

Bass, B. (1990). *Bass & Stodgill's handbook of leadership: Theory, research & managerial applications* (3rd ed.). New York, NY: The Free Press.

Byrne, D. (1998). *Complexity theory and the social sciences: An introduction.* New York: Routledge.

Clarke, O. (1995). *Oz Clarke's wine atlas: Wines & wine regions of the world.* New York: Little, Brown and Company.

Cooke-Davies, T., Cicmil, S., Crawford, L. & Richardson, K. (2007). We're not in Kansas anymore, Toto: Mapping the strange landscape of complexity theory, and its relationship to project management. *Project Management Journal, 38*(2), 50–61.

Duarte, D., & Snyder, N. (2006). *Mastering virtual teams: Strategies, tools, and techniques that succeed* (3rd ed.). San Francisco: Jossey-Bass.

Krajewski, L. & Ritzman, L. (1996). *Operations management, strategy and analysis* (4th ed.). Reading, MA: Addison-Wesley Publishing Company.

Leban, W. (2003). The relationship between leader behavior and emotional intelligence of the project manager and the success of complex projects (UMI No. 3092853).

Project Management Institute (Ed.). (2008). *A Guide to the Project Management Body of Knowledge—Fourth Edition*. Newtown Square, PA: PMI.

Project Management Institute (Ed.). (2009). *Practice standard project risk management*. Newtown Square, PA: PMI.

Reutzel, D. R. & Cooter, R. B. (2004). *Teaching children to read: Putting the pieces together*. (4th ed.) Upper Saddle River, NJ: Prentice Hall.

Singh, H. & Singh, A. (2002). Principles of complexity and chaos theory in project execution: A new approach to management. *Cost Engineering*, *44*(12), 23.

Standish Group (2009). Standish Chaos Report.

Titcomb, T. (1998). *Chaos and complexity theory*. Landisville, PA: ASTD.

Winter, M. & Smith, C. (2006). *Rethinking project management*. London, UK: Engineering and Physical Sciences Research Council.

Chapter 3

Virtual Leadership and Complexity

Figure 3.1 Water in the desert is an important part of life. Note how the plants closest to the water are green while plants just a few feet away appear brown.

Leaders who operate in the virtual environment have the most to gain from complexity. Complexity is about offering explanations for many social systems without limiting the explanation to conventional single variable understanding. Too often people want an explanation for all matters within a project in order to replicate this solution at a future time. What one must understand is that complexity is not about creating a single style solution. Complexity is about accepting that people operate in situations where there are multiple solutions. As society enters into leaner times, the role of the leader is also changing. The leader can no longer be omnipresent, as was possible in the past. Technology has done much to allow for leaders to better monitor larger numbers of people. Leaders must start to accept that their role is no longer one of the architect or engineer, but is now one of gardener or farmer of the individuals under their charge. As groups and teams continue to grow into more social tribes, the leader must understand that they cannot continue to operate as they have done in the past. Individuals must become more flexible and versed in directing the currents of their organization by working together within the environment rather than trying to reshape the environment into something that is rigid and inflexible, which is not in harmony with the environment.

WHAT ARE VIRTUAL PROJECTS AND WHY ARE THEY IMPORTANT?

Virtual organizations exist today, due to increased economic pressures of business and the desire to hire the best talent available. Moreover, the changing needs of flexible organizations continue to fuel the boom of virtual organizations and virtual teams. Virtual teams are clearly beneficial in numerous applications, and brick-and-mortar institutions may, in our lifetime, go the way of the dodo bird. Virtuality is becoming as common and ubiquitous as cable television. Organizations of the future will need to justify the need to imprison their workforce at a particular location when they can utilize fewer resources and expect more productivity in return by using virtual teams (Duarte & Snyder, 2006).

Virtuality is here to stay. Organizations can both embrace this new way of thinking and start applying new technologies to solve the problems of the future, or an organization can resist change and face the consequences of the future. Virtual teams and organizations offer a different look at management practices. The social world is in continuous motion, and the only thing that organizations can consider constant is change. Managers of virtual teams and virtual organizations must be aware of their evolving role, as well as the demands that will be placed upon them given the new circumstances of work. No longer can old methods be applied without serious implications of renewal, trust, motivation, and economics being considered. It is clear that the employees of the future will be knowledge workers tied loosely together toward a shared organizational goal. This new employee will make the ways of thinking in the past no longer applicable to the challenges of the future. Managers must break away from the past, and embrace the new structures of the dynamic future. Virtual teams are no longer an option for business; they are an expectation.

What defines a successful virtual project? As with traditional projects, the success factors need to be defined during the planning stages of the project. Success factors can be defined in terms of schedule, cost, quality, and other intangible factors. The project manager should develop a strategic plan to ensure that the success factors are built into the project.

A survey of the available literature on the subject makes clear that success rests upon four elements more than any other fundamentals reviewed. The successful virtual project managers must be prepared proactively to manage the team process to avert potential conflict; they must be versed in techniques to handle conflict when it does arise; they must utilize techniques to build trust in a dispersed team; and they must use appropriate leadership strategies.

The successful virtual project manager should also understand the differences in communication and technology needs within the dispersed project team. Anecdotal evidence suggests that project managers who create a "shadow" project plan for the virtual aspect of the project have succeeded more often. The virtual project plan should address the extra needs for communication, the need to standardize technology, and the need to energize the team.

Virtual teams are difficult for an organization to embrace, because they are perceived to be under less control than teams that are physically proximate. Companies that do not embrace virtual teams will find themselves at a competitive disadvantage within their industry. Virtual teams create greater value than physically proximate teams since it limits membership and increases costs of keeping the teams together. The successful virtual project manager understands that the virtual project actually has to have more processes and procedures in place in order to maintain control. The project manager understands that open and direct communication establishes trust, which academic research has demonstrated is essential to the successful virtual project.

Virtual project success is not guaranteed even when advanced technology is utilized. Research has shown that over 50 percent of projects came in up to 190 percent over budget and up to 220 percent late (Standish, 2009). In order to increase the odds of success, this research offers several winning strategies to the project manager to complete on time and within budget.

Practical Tip: There are many books out about virtual teams and the virtual environment. Consider reading one or more of those books to expand your knowledge about virtual organizations.

WHY COMPLEXITY IS A NECESSARY TOOL OF VIRTUAL LEADERSHIP

Complexity as it applies to project management is a new phenomenon. Aritua et al. (2009), Cooke-Davies et al. (2007), McKinnie (2007), and Jaafari (2003) demonstrated that complex projects were nonlinear, were self-organizing, were adaptable, used feedback, and relied on interrelations and interdependence with other organizations or environmental elements. These elements are part of complexity theory. According to McKinnie, the project managers interviewed acknowledged the presence of complexity but did not know how to manage or predict the chaos. Furthermore, McKinnie's study found that the *PMBOK*® (PMI, 2008) "was basically a document of definitions and methodologies. It did not incorporate a great deal of theory explaining project behavior" (p. 120).

Singh and Singh (2002) noted that projects traditionally exhibit the following complexity behavior:

> "uncontrollable behavior of the cost performance index; . . .
> Inability to follow schedule or the inability to explain a schedule in real terms; . . .
> Unpredictability of organizational and individual behavior; . . .
> Irrationality of teams; . . .
> Misbehavior of teams; . . .
> Failure of quality." (p. 24)

The first part of Figure 3.2 describes the disorganization that Singh and Singh (2002) explained as projects traditional elements. The second part describes how the ideal project is transformed from chaos into order. The *PMBOK*® offers various organizational techniques and ideas to assist a project manager, but it falls short when recognizing that projects consist of people and people do not always operate in an organized manner. The only offering the *PMBOK*® has in this area is in Appendix G, which has already been discussed as woefully inadequate and out of date with regard to new developments and complex virtual projects. Recently, complexity theory has offered a glimpse at how small groups operate effectively; however, even that research has been limited (Jaafari, 2003).

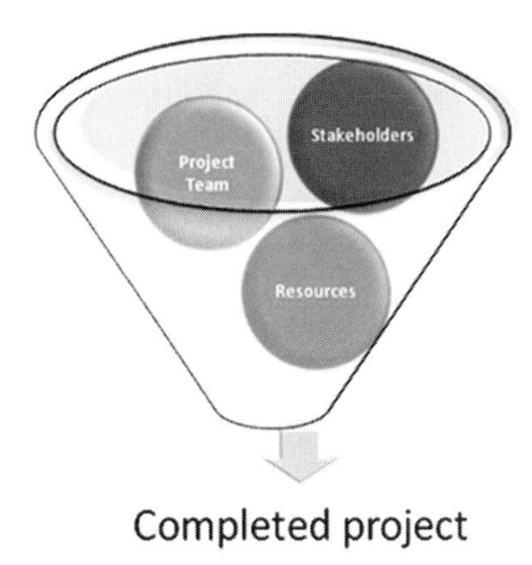

Figure 3.2 The funnel describes how a project is a blending of various elements rather than a linear relationship between elements.

Application of Complexity Theory to Virtual Teams

Nowadays society lends itself to complexity theory. Societies in general, and particularly societies that are adept at technology, tend to seek knowledge (Jaafari, 2003). When we do not know something or seek to verify information, we jump on the Internet. It may be as simple as using a search engine, going to a chat room, or blogging, or as intense as doing online academic research. Since knowledge is readily available, the complex project needs a different leader (Jaafari, 2003). This new type of virtual leader must not only be skilled and proficient to develop trust with his or her constituency, but must also embody and guide their network toward the goals of the project.

According to Jaafari's (2003) studies, which were supported by Leban's (2003) studies, transformational leadership was the style best suited for the complex project. These research studies suggested that project management certifications and strong-armed processes and procedures did not help the complex project. Hence, the virtual project manager needs to embrace the chaos and change that is inherent on the virtual project (Overman & Loraine, 1994). Even as early as 1994, Overman and Loraine realized that complexity theory needed to be applied to complex projects. Instead of trying to impose order on the "butterfly effects" as they occur, the project manager needs to embrace them and understand them. Therefore, the project manager should not impose order on the change but should understand it and help it.

The ability of the project manager to encourage and accept change within the team on complex projects should be seen as encouraging creativity and growth (Singh & Singh, 2002). In contrast, the need to control and maintain onerous processes and procedures encourages the *as is* on a project, which research suggests does not work on complex projects (Cooke-Davies, Cicmil, Crawford, & Richardson, 2007; Singh & Singh, 2002; Overman & Loraine, 1994). Singh and Singh suggest that the project manager should focus the team leaders on "dynamic instability" (2002, p. 31). Essentially, the project manager should encourage the teams to look at other alternatives.

WHAT MAKES VIRTUAL LEADERSHIP WORK?

Communication and leadership will make virtual projects successful. These two elements are necessary for success. Each requires intervention by the project manager, but also requires reciprocal support from team members and other affected groups.

Virtual leadership will work because of communication and trust. The nature of virtual projects requires additional communication in order to ensure that all elements are understood. Too often a lack of communication causes concern and consternation by team members—communication of the goals, communication of the project, communication of the achievements, and communication to everyone involved. Studies have shown that communication is required for any successful virtual team. The better the communication, the more likely the virtual team will be successful. Furthermore, the better the communication of the leader, the more likely the project will be successful (Duarte & Snyder, 2006; Lipnack & Stamps, 2000).

The second element is leadership. A leader must be continually assessing the project and must make sure that all the participants are aware of progress—the great motivator (Godin, 2008). Progress helps identify problem areas and areas of achievement. A leader must make sure all team members have the appropriate training and support for success.

Leaders must ensure that all team members are treated fairly. If a leader identifies an individual who lacks training or access to necessary communication or collaborative technology, the leader must strive to address this issue immediately (Duarte & Snyder, 2006; Lipnack & Stamps, 2000). Allowing certain individuals access to training or technology while disallowing others will create a rift in the team that will disrupt productivity. Leaders must maintain technology at a similar level to avoid the perception of favoritism. Although it may not be true, allowing advanced technology to certain individuals while disallowing the same technology to other individuals can potentially be negatively perceived as favoritism or worse, nepotism (Lipnack & Stamps, 2000)

Leaders must identify and classify their team members into a hierarchy that can support the team goals but also can be implicitly understood. Establishing and communicating the criteria for the hierarchy will often assist in solidifying the team's structure. A good leader should review with his or her team members to better understand their comfort and competence level in virtual teams. The leaders should also review the team structure to make sure it still assisting the team toward the goal.

Understanding the experience level of each team member will assist in the leader allocating time with each team member. Team members experienced in the virtual environment will need less direction and supervision than will those who are new to the virtual environment (Duarte & Snyder, 2006). This does not mean leaders can allow seasoned members to work on their own, but a good team leader can better allocate time when he or she understands the competence level of each team member. More than in traditional teams, virtual leaders must lead and communicate.

> Practical Tip: It is critical not only to communicate in a virtual team but to gain acknowledgment. One way to ensure that a message is received properly is to follow up after any decision has been reached. If the decision was done on the phone, follow up with an e-mail to document the decision. If the decision was sent via e-mail, follow up with a phone call to make sure that the message was understood. This will improve communication by ensuring that everyone understands the progress and requirements of the project.

How Leaders Can Help Make a Virtual Team Successful

Leadership and competence are the most important attributes of a successful virtual team. Team leaders are pivotal on a virtual team, because leadership is one of the major components of a successful virtual team. While leadership is important, communication and collaboration is even more important.

Leaders who communicate and collaborate rather than forcing agendas or issuing orders are more successful (Godin, 2008). Leaders must not only take the team to a new level, but must collaborate with them to achieve mutual success. Project success is critical; there is significant data that supports individual success to be equally important. Being part of a successful project is good for one's career, but being personally successful is invigorating—for example, being part of a successful marketing campaign that increases distribution and use of a particular product. However, individuals might feel personally different if the product turns out to have harmful side effects for those who use it. Success is important, but individual fulfillment becomes more essential for a project. Virtual project leaders who learn to bake individual pride in with project goals will find individuals putting forth greater effort toward the success of a project.

TRANSFORMATIONAL LEADERSHIP IN VIRTUAL TEAMS

There are several known elements for effective virtual leadership. These elements include transformational leadership, using an appropriate leadership style and tone, sufficient experience of the team leader, and the use of standard processes. Transformational leadership is one of the effective styles of leadership for virtual teams. Transformational leaders set high expectations for the team members and model appropriate team behavior. In addition, the transformational leader is able to lead a team with disparate views and ideas toward a common objective. These elements relate to the transformational leadership style, and hence the Project Manager (PM) should consider this style for use in the virtual environment.

Appropriate leadership is also important. The PM should establish metrics that measure individual and team performance. These metrics should institute high expectations to encourage the extra effort required to overcome the communication hurdles. As the team becomes familiar with the team expectations and the performance metrics, the amount of undesirable conflict should be reduced. One needs to keep the team focused on the long- and short-term goals. It is important that these metrics be tied to important milestones and should involve appropriate incentives and celebrations. Without this feedback to the team, the PM will appear to be a taskmaster bent upon increasing efficiency rather than a leader trying to achieve team success. Since virtual teams are dependent upon technology for communication, the PM must make sure that appropriate technology is available and operational. Access to comparable electronic communication technology is important, yet research does not support the idea that state-of-the-art equipment is imperative for success.

Team leaders with virtual experience were also more successful, particularly those having experience working across organizational and cultural boundaries. Successful project managers are those who adapt best practices and competencies to their own projects based upon their past experience. The successful PM understands the needs of the members, and adapts rules and regulations to increase the relationship and trust among the team between leader and members. A successful virtual leader must be competent and adept at the following: developing and transitioning team members, developing and adapting organizational processes to meet the team's needs, allowing leadership to transition when appropriate, and ensuring the team receives appropriate training for virtual communications and technology.

The virtual environment can be confusing to first-timers, as well as some veterans. Because of this potential confusion, the PM should adopt standard processes within the team, which are agreed upon from the start of the project. Soft processes for the virtual environment should be agreed upon, including items such as conflict resolution and communication. The project manager and the team members should also be competent and understand the importance behind the standardization. Adopting a predictable approach will assist in avoiding misunderstandings about process as a reason for conflict.

All of these factors are important in virtual teams. Wise leaders would familiarize themselves with these techniques in order to be more successful. Even experienced virtual leaders need to understand all of these basic components so they can be more successful in the future.

> Practical Tip: A leader should be transformational in their behavior by offering a vision of the future and a plan to reach that new objective. For example, the project manager will celebrate milestones by sending congratulatory e-cards to project members. However, the project manager will also ensure that he or she understands the work hours of the team and will, if necessary, enforce a policy of time off.

HOW LEADERS CAN IMPLEMENT COMPLEXITY TO HELP MAKE A VIRTUAL TEAM SUCCESSFUL

Leaders must learn to understand that their virtual team's success is based upon the team's ability to become more than just a collection of individuals with one common purpose. Virtual teams can become much more than their component parts and individual elements. Complexity can become a part of these teams in order to propel them toward greater success. Although the moniker of teams has been around for a long time, there is often a misunderstanding of their purpose. Teams are not just groups of people who happen to work on the same project, but a true team is a community of individuals determined toward success. A common goal is not a bad thing, but there is more to a team than the common goal.

The first step in this process is to forget all you believe you know about leadership and start to consider that no one leader can do everything in a project, nor can they totally understand every aspect of any project that they will ever be a part of. Consider that no aspect of human knowledge can be concentrated into any book, degree program, or training program that will ever offer even a fraction of a percent of the available knowledge on any given topic. Individuals need to embrace the concept; no project manager can ever hope to have all the answers regarding any project that they are a part of. What they can hope to achieve is to offer good recommendations.

In order even to consider utilizing complexity, inventory the team and take a hard look at what the team is all about. If the group is just about the project and they are only together for that purpose, then it is unlikely that complexity or any other team process will garner the type of productivity that project managers crave. The team (or tribe) must be motivated toward success. Team greatness is more than just spreadsheets, timelines, and milestones. Consider whether the team operates as a family, as a neighborhood, or as a tribe. Project managers must consider if the team is all about working together and being a part of a greater whole (Godin, 2008, Lipnack & Stamps, 2000).

This team camaraderie is often easy to spot when individuals are actively engaged and the project is something that people are passionate about. Team members must enjoy what they are doing. They must see the future of the project and want to be a part of it, regardless of the outcome.

Team members should actively support one another. When team members care enough about the success and well being of others on the team, there is a special bond that grows between individuals (Godin, 2008). When team members are connected by more than the strings of a project, individuals begin to collectively come together in a way that allows them to get past petty situations. Team members concerned about others take on a new persona where they actively seek to create better people rather than just looking for individual productivity gains. Great projects come from great people and great people understand that a group of like-minded and motivated individuals can create more than just a group of individuals who share a common goal. Team members who do not share this kind of bond are less likely to assist one another in times of need, which can be a problem for any project that is coming upon a deadline.

Another good indicator is when the team pulls together in times of crisis. The very common reaction in any crisis situation is to shift blame as quickly and effectively as possible. Entire careers of individuals are all about just shifting blame. Many organizations actually reward individuals who can shift blame most effectively. This is very unfortunate, but at least it is a good indication the team needs more work before starting to steer it toward complexity. When one can find these elements of passion, support, and coming together as part of the team, then one should consider injecting elements of complexity into the organization to make it more effective and productive. If a project team is not at this point, or is only partially there, then it is up to the project manager to move the team closer to this higher level of cohesion. Certain techniques can help move a team toward this level of cohesion; these will be addressed in later chapters. This is purposefully done in order to offer a better view of the final picture of how complexity can benefit an organization rather than obscure the goals with additional material to bring a team to a point to be ready to accept complexity.

The second step of the process is to learn how to integrate complexity as part of that strategy to offer good recommendations. This is achieved by exploiting successes in order to leverage the uncertain future (Drucker, 2001). A project manager needs to learn how to build upon prior successes in a way that allows a leader to do more in the future. Often times a great success can help move an organization forward. Sometimes a success stands out so much that future groups are compelled to try to achieve as much, if not more. This example is often seen in sports teams dynasties, where an organization is so successful at winning that it attracts more talent in that area, which in turn spurs greater success.

In sports, teams are constantly changing. Team members will come and go, often at the whim of the market. At the college level, team members graduate and new team members enter the school. So, how can one have the logical expectation that the team that follows a great team will be equally good (or better). If the talent moves on, then the project or program should suffer. Or does it? If the sums of the parts are less than before, then how is it possible to achieve more? Therein lies one of the operational advantages of complexity—the sum of the parts can be more (or less) than the calculated total. This depends upon how the people interact. People, unlike a common machine, either work together or they do not. In rare examples, a worn part might continue to function at a reduced rate, but the parts can never exceed their specifications. A team of people can. That is one of the great examples to consider regarding productivity and why one should integrate complexity as part of a unified strategy.

The final step is to foster new neighborhoods of communal learning that operate with complexity. The concept is not new to encourage people to operate in neighborhoods, but the secret lies within how enthusiastic the people in the neighborhood are. Teams with enthusiasm will be more successful than teams without enthusiasm. For example, if there are two classrooms and the same teacher, and one classroom has twenty students and the other classroom has fifty students, which classroom will perform better at the end of the term? The expectation is that the classroom with twenty students would do better.

Now, take the same two class sizes and the room of twenty students are forced to take the course as punishment (think driving school) and the classroom with fifty students are all enthusiastic and want to learn how to drive (think about a first time drivers education class). The class of fifty students will probably end up learning more (Godin, 2008).

The goal of a great leader is to create enthusiasm about the project. A leader can coach, a leader can guide, a leader can paint the vision of the future, but if more time can be spent to create a neighborhood of learning (ideally about a project), the team can learn and achieve more. This can be achieved with complexity by following some (or all) of these small ideas that can lead to creating communities of communal learning.

1. Publish the purpose of the project. Make it short but avoid anything to do with the financial implications. Make the purpose statement why the project is great for everyone. Also, make it as short as possible. Statements like "Remember the Alamo!" still carry meaning to people that cannot even find the location of the Alamo on a map.

2. Connect with everyone. The leader must find a way to connect with people. The more connections, the more people that associate with the project. Association is often the key to success. Humorous association often goes a long way as well. In one particular project, there was an issue with vibration on a vessel. The vessels were dubbed the "Shake, rattle and roll." With such a catching name, soon everyone would be asking about

the project and if the problem was solved. The problem was not about knowing what to do, but to find a way to get it done economically.

The cost to rebuild the area would be expensive and the project was already in financial trouble. The project team kept putting off the rebuild in hopes of finding a better solution. Eventually, the project vibration problem became so visible in the organization that several new engineers were brought into the project to help with the problem.

In the end, the team of engineers came up with an innovative and never tried before solution, which not only solved the problem but created a new market for the material that was used. The engineers used spray insulation in the air gaps between the walls, floors and ceiling that surrounded the source of the vibration—in essence, isolating the problem to only one room. The spray insulation was not rated for sound dampening because the usual application was a thin layer (which was sufficient to insulate). The advantage was that without tearing apart any walls, six inches of insulation could be pumped to isolate the source of the noise and vibration. So, the problem was solved with minimal costs because the room did not need to be ripped apart and costly sound dampening material was not necessary to be installed throughout the area. All that was required was spray insulation and one person to direct the hose of nozzle of the spray insulation into a small hole that would be drilled into the wall.

3. Get the team to easily connect with one another. The more the team can work together, the more those on the team will solve problems which will keep the project from slowing down. Often times, making sure that a team has face-to-face interactions at the beginning will help people connect (Duarte & Snyder, 2006). A virtual team often feels disconnected, so a leader should take steps to bring them together whenever possible. Connections make a difference.

WINNING STRATEGIES FOR VIRTUAL PROJECT MANAGERS

Research into the winning strategies for virtual project managers has shown several winning strategies available to virtual project managers. In order to understand the strategies of successful project managers, it is necessary to study the elements of successful projects. During the study of successful projects, four strategies have been identified as the best for a project manager to utilize. The research found that three of the most successful strategies for project managers related to the project managers' competency and how they led. The fourth was related to technology

(Duarte & Snyder, 2006). It is important to understand that a combination of these strategies is the best approach, and virtual project managers should do their best to combine these strategies.

Leadership is a major component of a successful virtual team. Teams that are led well feel that this assists in the completion of the project as well as in the success of the time. Successful project managers are those with expertise in virtual projects, those able to mobilize internal support and resources for a project, and those who set high expectations for team members (Duarte & Snyder, 2006; Lipnack & Stamps, 2000). Research has shown that successful project managers in the virtual environment are most tightly correlated with the following attributes: virtual project managers gained support of stakeholders, had experience with dealing with cross-cultural teams, established high expectations, and technology was consistent and available to all virtual team members.

There was also a correlation between the virtual team and technology. Technology is important and team members require a high commitment of time when training with technology. Technology without training does not have the same level of correlation. A project manager must make sure to involve people with the technology as well as to ensure that they are trained and comfortable with the available tools (Duarte & Snyder, 2006; Lipnack & Stamps, 2000).

> Practical Tip: Everyone wants to be on a winning team and not on a whining team. Consider what can be done to shape the internal and external perception of the team. Do whatever you can to make every team that you are a part of a winning team.

Leadership is critical for the success of virtual teams. Leaders must support the team as well as help them move toward the desired goal. However, task leadership is not sufficient for success in the virtual environment. Leaders who only communicated tasks and timelines were not as successful as those who truly led the team.

Leaders not only must help the team toward the new goal, but they must identify and garner support from the extended network of experts that support the group (Godin, 2008). In order to make the project successful, the PM must identify these stakeholders and gain their support. One successful strategy involves creating a stakeholder matrix that outlines the level of participation, roles, and contact information so that all team members understand the resources and requirements expected of the team. Not only does this communicate to team members the listing of stakeholders, but it also helps them understand all the important participants in any project. A team that becomes a tool of positive communication to all stakeholders can be a powerful tool in project success. Furthermore, successful virtual project managers are skilled at gaining the support of customers and stakeholders. The project leader must be able not only to rally the troops, but to rally support for the project from all areas of the organization.

> Practical Tip: Build a stakeholder matrix and communicate this information to all team members. Ask team members if they ever pass along information regarding the project to these stakeholders. Create specific e-mail distribution lists to make sure that all stakeholders are copied on all important correspondence. Also consider creating a specific Intranet or Web site to support the project. The more communication done in such a venue, the better the project can be monitored by others. Creating positive marketing of a project will help ensure its success.

The successful PM understands the needs of the members and adapts rules and regulations to increase the relationship and trust among the members, and between leader and members. A best practice for a virtual project leader is to adapt the competencies of different cultures. Trying to go with the flow, rather than to fight the current, is a wise move for a project manager. Understanding and then leveraging each team member's strengths is critical. Team members will be motivated to contribute positively to the group and will be proud to have the opportunity to contribute to the group.

Other successful leadership strategies for projects are developing and transitioning team members, developing and adapting organizational processes to meet the team's needs, allowing leadership to transition when appropriate, and ensuring the team receives appropriate training for virtual communications and technology and skill sets.

A PM must consider whether he or she has the ability to develop and transition team members as the project requires. All projects have a degree of transience because team members will shift in and out of the project as the various milestones are reached. Given that all team members will not be active throughout the project, a PM must that expect team members will enter and exit the project smoothly (Duarte & Snyder, 2006). Being able to transition these team members rapidly and effectively will assist in making the project a success. This may not seem directly related to the project timeline, but it will affect the overall morale, trust, and feelings of the team members.

A PM should review the changes required by the project and then map out the new organizational processes that will meet these new requirements.

Leadership transition is important to the success of a virtual project. The PM will not always be the person best suited to lead the team at all times. At times it would be better to allow someone else, perhaps with more technical ability, or an influential stakeholder, to lead the team on an interim basis.

Of the different leadership styles available, the experts most often associated the transformational leadership style with success. This does not mean that other leadership styles are not successful; it was just observed that transformational seemed to correlate greater with success. One possible reason for this success is for a PM to be transformational; the PM must understand the needs of the team members, and to pass on knowledge, which correlates to project success.

The data support the concept that technology is necessary for a successful team. However, state-of-the-art technology is not necessary for success (Duarte & Snyder, 2006). Access to comparable electronic communication technology is important, yet no research was found to support the need to have state-of-the-art equipment.

CREATING METRICS TO TRACK THE SUCCESS OF COMPLEXITY

The PM should establish metrics that measure team performance. These metrics should institute high expectations to encourage the extra effort required to overcome the communication hurdles. As the team becomes familiar with team expectations and performance metrics, the amount of undesirable conflict should be reduced. It is important that these metrics be tied to important milestones and involve appropriate incentives and celebrations. Without this feedback to the team, the PM will appear to be a taskmaster bent upon increasing efficiency rather than a leader trying to achieve success.

High expectations should always be defined as stretch goals for the team. In any organization there is an expectation and then there is the expectation for an exceptional employee. Create clear stretch goals that outline not only the expectation but also what would be an exceptional result. For example, if a project task is to be completed on August 1 (expected result), a stretch goal for that same task would be to have it completed by July 7. Clearly if this example was a predecessor to more important goals then completing the task three weeks early would be of great benefit to the project team. Hence, the project manager is setting an expected goal per the project timeline, but also setting a higher goal for team members to strive for. This can be done at every level and for every task. It is not unrealistic for a project manager to create an *ideal* timeline for the project, where tasks and milestones are completed early and then there is more time for testing and evaluation at the end—or even complete the project early and deliver the product to the customer early.

One often sees this kind of *dream* timeline created when there is an early completion bonus because then the team is well motivated toward earlier results; however, there is no reason that this should not be done for every project. Too often a project manager forgets that the timeline is both a tool and a measurement of success. Leveraging this to the advantage of the team is often sufficient to make superior results.

Some individuals might feel that putting additional stress on a team is going to result in conflict as project members scramble and struggle with resources and unrealistic deadlines. What a project manager forgets is that first, adversity is the mother of invention, and second, that who knows what is truly an unrealistic goal. With regard to creating adversity to stimulate creativity, this is something that happens all the time.

First, creativity happens when people have a goal that needs to be achieved, but they lack sufficient time to complete the goal. Offering stimuli to help creativity is often the best way not only to get the project done faster, but to make it fun. Individuals will then be challenged to create new tools that can complete tasks more efficiently. If team members have no reason to perform tasks in the same manner as they always have, then a project manager can be sure that they will take the route of least resistance. If one offers a new challenging goal, then the team is more motivated toward greater success.

Second, when one looks back on the creation of most timelines, these constructs are often based upon prior projects, experience, and expert guesstimates. So some kind of slack should always be built into any timeline. Even if a project is based upon being similar to an earlier project, there should be better ways to complete the same tasks. An extreme example of this can be seen in the building of Victory ships during World War II in the United States. There is no doubt that the building of any vessel is a lengthy and complicated process and is highly labor intensive. Prior to World War II, ships would take months and months to construct and there was no reason to speed up the process. However, by the close of World War II, the United States was able to complete the construction of entire ships in only a few days. The external pressure of winning the war stimulated everyone involved in the process to try new things that were never even conceived in the past. Some would say that improved technology aided in this shift; however, greater achievement was accomplished by the application of technology rather than new technology. After all, there was not a new ship-building machine created to build ships in days—the success came from the people involved in the process becoming better organized and efficient, and from the development of standardized processes that would lead to success.

Communication between the trades increased and conflict was reduced, not increased. The looming concern about the war developed a new process that bonded everyone toward the same goal. Rivalries were reduced and even though resources were short, teams learned to scrounge and do more with less. Project managers will certainly be concerned about other possible effects, but they must learn to stretch themselves in order to have others achieve those greater goals.

CHAPTER SUMMARY

Virtual teams are all about having people working in the same way as if they were working in adjacent cubes. There is no easy way to force people to want to work together, so ultimately the project manager must assist in getting people to want to work together. The more the project manager can get people on the same page, the more powerful the project can be. Imagine the difference between thousands of people walking the same direction in unison (a parade) and thousands of people all walking in their own particular way while doing their own thing (a riot). The difference is staggering. Consider how impressive both a parade and a riot can be. Yet one is a force to improve society while the other is a force that can bring down society.

REFERENCES

Aritua, B., Smith, N., & Bower, D. (2009). Construction client multi-projects: A complex adaptive systems perspective. *International Journal of Project Management*, *27*, 72–79.

Cooke-Davies, T., Cicmil, S., Crawford, L. & Richardson, K. (2007). We're not in Kansas anymore, Toto: Mapping the strange landscape of complexity theory, and its relationship to project management. *Project Management Journal*, *38*(2), 50–61.

Drucker, P. (2001). *Management challenges for the 21st century*. NY, NY: HarperBusiness

Duarte, D. & Snyder, N. (2006). *Mastering virtual teams: Strategies, Tools, and Techniques That Succeed*. San Francisco: Jossey-Bass

Godin, S. (2008). *Tribes*. NY, NY: Penguin Group

Jaafari, A. (2003). Project management in the age of complexity and change. *Project Management Journal*, *34*(4), 47–57.

Leban, W. (2003). The relationship between leader behavior and emotional intelligence of the project manager and the success of complex projects (UMI No. 3092853).

Lipnack. J. & Stamps, J. (2000). *Virtual teams: People working across boundaries with technology* (2nd ed.). NY, NY: John Wiley & Sons

McKinnie, R. (2007). The application of complexity theory to the field of project management. (UMI No. 3283983).

Overman, E. & Loraine, D. (1994). Information for control: Another management proverb. *Public Administration Review*, *54*(2), 193–196.

Project Management Institute (Ed.). (2008). *A Guide to the Project Management Body of Knowledge—Fourth Edition*. Newtown Square, PA: PMI.

Project Management Institute (Ed.). (2009). *Practice standard project risk management*. Newtown Square, PA: PMI.

Singh, H. & A. Singh (2002). Principles of complexity and chaos theory in project execution: A new approach to management. *Cost Engineering*, *44*(12), 23–33.

Standish Chaos Report (2009). Standish Group.

Chapter 4

Successful Virtual Projects

Figure 4.1 The channel between the boats is a path of water that leads to the ocean. This is the same for virtual projects, where one must start in the calm of planning before entering the potentially torrential sea of the project.

Successful virtual projects are much like the quandary of the chicken and the egg. Did successful virtual teams come first and then came successful virtual projects, or did a team's success create the first successful virtual project? Although interesting, this quandary helps illustrate that virtual projects are more difficult to achieve than one would believe. Regardless of which came first, there is no doubt the leadership of the team is always at the heart of the success. Teams that do well are, regardless of the environment, teams that are led well. The problem is indentifying what makes for a successful leader. Furthermore, what makes for a consistent successful leader? Many books have been written about projects that were successful because the leader was good, but it is clear that there is no one perfect way to lead any team. Leaders must continue to adapt and manage the expectations of team members, shareholders, and stakeholders.

INTRODUCTION TO SUCCESSFUL VIRTUAL PROJECTS

A virtual team is a group of individuals who work across space, time, and organizational boundaries who are brought together to perform interdependent tasks, while occupying geographically different locations, united by communications technology and a common purpose. Or, more succinctly, a virtual team is a boundaryless networked organization assembled to perform a task where the team is coordinated though trust and shared communications. The virtual environment combines elements of virtual teams and virtual organizations; hence, the virtual environment is the boundaryless, networked work setting that binds geographically distributed individuals by communications technology.

Organizations must become chameleons in order to survive in business today. This creative metaphor offers a view of the proper relationship of a future organization. The organization of the future must morph like the chameleon. Its color, shape, size, and appearance will alter as its environment and the resulting demands will evolve. This distinctive metaphor describes the elements of virtual organizations. The organization is flexible and adapts to the environment rather than the normal human tendency to alter the environment to adapt to the organization.

Another metaphor for the virtual organization is the corporate condominium. Imagining that the companies of the future are condominiums is an excellent metaphor for this new model of business. Virtual organizations are more than just temporary brain trusts; they are organizations of greater ownership than those that exist just to work in an organization. Real ownership implies a sharing of the wealth and a sharing of the information of the organization. The sharing of knowledge can happen in a virtual organization where every person can feel a valuable member of the whole. Virtual organizations are going to be very different from the organizations of the past.

Virtual organizations exist today, due to increased economic pressures of business, and the desire to hire the best talent available. Moreover, the changing needs of flexible organizations continue to fuel the boom of virtual organizations and virtual teams. Virtual teams are clearly beneficial in numerous applications, and brick-and-mortar institutions may, in our lifetime, go the way of the dodo bird. Virtuality is becoming as common and ubiquitous as cable television. Organizations of the future will need to justify why there is a need to imprison their workforce at a particular location when they can utilize fewer resources and expect more productivity in return by using virtual teams.

Virtuality is here to stay, and organizations can both embrace this new way of thinking and start applying new technologies to solve the problems of the future, or an organization can resist change and face the consequences of the future. Virtual teams and organizations offer a fresh look at management practices of organizations. The social world is in continuous motion, and the only thing that organizations can consider constant is change. Managers of virtual teams and virtual organizations must be aware of their evolving role, as well as the demands that will be placed upon them given the new circumstances of work. No longer can old methods be applied without serious implications of renewal, trust, motivation, and economics being considered. It is clear that the employees of the future will be knowledge workers tied loosely together toward a shared organizational goal. This new virtual employee will make the ways of thinking in the past no longer applicable to the challenges of the future. Managers must break away from the past and embrace the new structures of the dynamic future. Virtual teams are no longer an option for business; they are an expectation.

Surveying the available literature on the subject makes clear that success rests upon four elements more than any other fundamentals reviewed. Successful virtual project managers must be prepared proactively to manage the team process to avert potential conflict; they must be versed in techniques to handle conflict when it does arise; they must utilize techniques to build trust in a dispersed team; and they must use appropriate leadership strategies.

The successful virtual project manager should also understand the differences in communication and technology needs within the dispersed project team. Anecdotal evidence suggests that those project managers create a "shadow" project plan for the virtual aspect of the project to have succeeded more often. The virtual project plan should address the extra needs for communication, the need to standardize technology, and the need to energize the team.

Virtual teams are difficult for an organization to embrace because they are perceived to be under less control than teams that are physically proximate (Duarte & Snyder, 2006). Companies that do not embrace virtual teams will find themselves at a competitive disadvantage within their industry. Virtual teams create greater value than physically proximate teams; the latter situation limits membership and increases costs of keeping the teams together. The successful virtual project manager understands the virtual project actually has to have more processes and procedures in place in order to maintain control. The project manager understands that open and direct communication establishes trust, which academic research has demonstrated is essential on the successful virtual project.

Virtual project success is not guaranteed even when advanced technology is utilized. Research has shown that over 50 percent of projects came in up to 190 percent over budget and up to 220 percent late (Standish, 2009). In order to increase the odds of success, this research is offering several winning strategies to the project manager to complete on time and within budget.

> Practical Tip: Virtual organizations can make some people uncomfortable. There is no water cooler time or other socialization of team members. To avoid void in the team experience, create virtual message boards to help team members reach out to one another for matters outside the project at hand.

TRUST IN THE VIRTUAL PROJECT

For a virtual organization to be successful, trust must be part of the foundation of the organization. Trust demands boundaries and learning. Trust requires bonding and leaders that have a certain touch to management. Trust includes such factors as competence, integrity, and a concern for others (stakeholders). These elements of trust are necessary to make an organization successful. Trust must be visible at all levels of an organization in order to achieve greatness. No virtual organization can be successful, let alone effective, without the individuals involved having a high level of trust at all levels.

Successful virtual teams require face-to-face meetings initially to build camaraderie and trust among members of the team. The Project Manager (PM) who conducts himself or herself as a strategist has the ability to create an environment of trust. This becomes an environment conducive to making organizational changes. Trust can be created at the same time when the leader is building up the team to support the leader. Leaders who position themselves as the mechanism for change help generate this trust. This trust-building work enhances the success of virtual teams. Trust is essential to the virtual environment's success, and the more time the team spends together, the more trust increases among the participants.

The PM's leadership style benefits by promoting trust and collaboration in a faceless environment. Creative manners and communication help the PM establish trust. The effective PM establishes trust between himself or herself and each team member individually. A PM must also promote trust within the group. The PM must devote time to building relationships with each team member. The PM must learn how to interact with each team member and how to keep track of individual progress. A PM must be interested not only in the project, but in the person. This will make a difference when building trust. Individuals trust those who know them.

Trust does not end with the individual; the PM must build trust as part of a group. Group members must learn to trust one another and to trust the judgment of the group. In most cases, the more the team has time together, the better they will feel about one another. Give the team time to interact and to learn about one another. Trust is created when individuals respect one another. Trust must be visible at all levels of an organization in order to achieve greatness. No virtual organization can be successful, let alone effective, without the individuals involved having a high level of trust in one another

Leaders of virtual teams should avoid the common error of using just techniques that create trust for a group; rather the PM must embrace the actual ideas involved. For example, a team meeting will not build trust when people just report out on their progress. A staff meeting does nothing to build trust if trust is not on the agenda. If a PM wants to use this time to build trust, put trust at the top of the agenda and stick to it. Talk to the team about trust and have them talk to one another about what trust means to them. Communicate about how effective trust is (or is not) being built in the project. The results might surprise you once you have a group talking about trust. A PM must remember that building trust is not an either/or proposition. Trust must be built individually and it must be built as a group. Studies have clearly shown that without trust, a virtual team, and the project, is more likely to fail.

One study found that executives in eight major U.S. corporations agree that it is difficult to establish trust in a multicultural environment.

This lack of trust leads to companies establishing duplicate processes and procedures and redundant systems. This in turn results in many international companies having multiple parallel enterprises instead of one cohesive enterprise. If supervisors do not already trust and respect virtual employees, then they need to build trust and respect for virtual employees. Many managers need visual clues for interrelationships to be comfortable. This kind of feedback is important as people determine if their message is being received and accepted. A virtual PM needs to learn how to move past this managerial limitation and learn to rely upon other cues for security.

> Practical Tip: Trust is always earned. There are no fast solutions for trust. One cannot create trust by saying one thing and doing another—trust must be expressed in actions and words. Talk about trust in the organization and keep talking about it until you feel that trust exists at all levels.

> Practical Tip: Budget for face-to-face time between team members to facilitate communication. If that is not possible, schedule conference calls and other interaction between team members to make sure that everyone is apprised of the project.

Trust and Communication

Organizational communication is complex and important to the virtual team. Organizations are no longer visible, tangible places where managers can wander around to learn about what is going on. Management by walking around is no longer a viable option in the virtual world. Virtual workers need to learn to communicate better because there is less contact time for individuals. As technology improves and becomes less expensive to implement, virtual organizations will become the norm. People are not connected only by phone, fax, and e-mail. The continuous monitoring of team members that an organization of the past considered essential is no longer necessary. Work will be done in a manner where no one can see it happen. Individuals in organizations of the future will learn to make a few bold strokes and then pass the brush along to someone else, who will then add their own perspective to the growing work of art. To meet this new way of conducting business, a PM will need to increase the level of communication.

There are three views to communication. In the end, all of these styles of communication can be effective. A PM has to decide which one(s) they will use consistently and then use them. Communication builds trust when there is predictability. First, one can approach communication as evolving from the face-to-face communication necessary in hierarchical structures to a vehicle to convey culture through stories. Some managers feel that if they can tell a story that parallels the current situation with a positive outcome, they can help individuals deal with the current situation. In the past, managers would build this trust through direct interaction, but that option is no longer possible.

Face-to-face communication is quickly being supplanted by different methods of communication. Telephones and electronic mail have replaced face-to-face communication when passing along stories that support organizational lore. The spoken word continues to be important, and stories have become a method to explain culture and values of an organization. Furthermore, the written word in the form of e-mail has become even more important as managers find themselves communicating more via e-mail to their subordinates than ever before.

Second, communication and trust building can be seen as being directed by a series of gatekeepers, individuals who control the flow of communication to others. This view addresses the virtual relationship of individuals as mediators in a dynamic communication systems; a relationship of cause and effect. Gatekeepers in the virtual environment can either restrict information or mediate communication. Either form can be altruistic; however, these gatekeepers can be far more insidious when they are skilled politicians who control knowledge to maintain their expert status. A manager might feel more in control when using this method, but others will always see this form of restricted communication as negative. When information becomes a medium of exchange, it can create some relationships that are not founded in trust. The PM must use this kind of communication carefully.

Third, in contrast to these views about gatekeepers, others believe the virtual organization is not about the single steward whose actions control the organization's destiny. Leading a virtual organization is about a single individual who harnesses a personal and business network in order to achieve impressive results. Virtual teams require that leaders shift from a focus upon personal ability to a focus on group results. Individuals who are interested in leveraging the strengths associated with virtual work must increasingly follow this pattern of success through networking in the future.

This view of harnessing the group seems to offer the best use of resources, but if improperly handled it can lead to cyberchaos. Cyberchaos results from information overload where nothing can be done because individuals feel paralyzed by the data available. The PM must learn to offer information that results in real communication within an organization. Real communication is the only way to challenge this glut of information. Leaders must perform actions that support communication, while including the appropriate tone along with the message of the words. Effective distribution and comprehension of data to all levels of an organization are the essential elements of real communication.

Ultimately, contemporary organization must learn to communicate better in order to build trust and to ensure that their messages are received and comprehended. The communication distribution system has emerged as an important force in effective internal and external organizational communications. This is particularly important in virtual teams because of the potential impact.

> Practical Tip: Face-to-face communication is important to build trust. Sometimes trust is built up over time by proximity. An example of this phenomenon is the bond between neighbors. You may not know your neighbors, but seeing them and passing them in your daily routine will make you more likely to strike up conversation with them than you would with a total stranger.

Trust and Change

The important reason for building trust in any organization is the desire to enact change. Change requires planning, organizing, controlling, and leading. Planned change never happens in a vacuum. Managers that implement change must be prepared to lead and to stay the course. Leading for change also takes good followers. Leaders must not only know their followers, but must trust them and vice versa. So how does this lead to change?

Effective change comes when leadership plans it; however, the largest obstacle to change is always the resistance to new ideas. A leader must have the respect of followers and must remain steadfast whenever encountering resistance. Anyone can steer a ship when the seas are calm, but it takes a true leader to maintain a steady course when the seas are rough. To overcome resistance to change, the leader must understand certain realities associated with any change.

In order to discuss resistance to change, one must first identify the reasons for resistance. Everyone prefers to avoid change, and it is important to understand the social undercurrents of this phenomenon. It is natural to avoid jumping out of an airplane because a fall from that height would result in certain death, yet when an individual is skilled and has a parachute, this feat becomes far less insidious.

Several elements keep organizations from changing. They exist in all groups and are not restricted just to virtual teams, so it is important to understand these elements and be able to identify them so that they can be discussed. The only way to combat these is to bring them out in the open so they can be discussed by the organization.

Homeostasis: Continual change is not a natural condition of life; hence, human beings resist this state of change and prefer stability. The counter to this type of resistance is to sell the parties on the benefits of the change (O'Toole, 1996). One must explain the benefits as opposed to dwelling upon the shortcomings involved with the change. Do not be afraid to address the big concerns, as this will help to sell the change. Also, do not be afraid to present the possible outcome if the change is not completed. Sometimes change means the difference between the company surviving tough economic times and going under. One also needs to be careful to recognize when a person is willing to change. There is nothing more annoying than a person trying to sell to someone who already understands and believes in the change.

Individual Fear: Humans are always afraid of the unknown. There is no way to avoid this element because we all fear something. Again this is an opportunity for education. People fear the unknown, so through education people can learn to embrace the change faster. Everyone is a little nervous at the start of a new class. Fear is a natural reaction, so we must always contend with a certain level of fear. The more one can do to eliminate or diffuse that fear, the easier change becomes.

Selfish Self-Interest: The *what's in it for me* attitude best describes the challenge of self-interest. Often individuals want to disregard this type of resistance, but it is certainly a motivator for many people. If people do not see the direct benefit, it is hard for them to embrace the change. The solution to this form of resistance is to create a *win-win* solution. The leader has to make the change more enticing to people than resisting the change. This may mean changing the reward system or maybe even the performance appraisal system. Make whatever changes you can to make change attractive to all people involved. It is often also helpful to link people together so they may be successful in order to achieve a reward.

Cynicism: It is human nature to believe that planned change is doomed to failure, and we will naturally suspect anyone who is trying to change us (O'Toole, 1996). Cynicism is a common concern of change agents. They feel that leading change leaves the initiator open to criticism. One must learn not to be afraid of criticism. Rejection is just a sign of resistance to change and has nothing to do with the value of the idea. The only way to effectively counter cynicism is to strengthen those who believe in the change. Spending time trying to change a critic is not valuable. What is valuable is reinforcing your position and those who support it, by overwhelming the opposition with the facts about the change. Critics change their tune quickly when faced with serious and focused opposition. It is often good to approach critics in an open forum with the facts. It is harder for them to resist the change if they openly must explain their resistance. This method is good if the change can be shown as positive, but one must make sure to have all the details in order to avoid falling victim to the cynic. A good cynic will have good and often logical reasons for the resistance and if one is not able to counter their response, it is not a good idea to challenge them openly. Cynicism can spread quickly if a group does not see the individual benefit.

Group Fear: Groups often act contrary to reason and enlightened self-interest. Groupthink can be dangerous and many great ideas have been met with violent opposition from mediocre minds. Great ideas are not always embraced by groups, even when those groups may be the direct beneficiaries of the change. Groupthink is the bane of all change. People will naturally huddle together in order to avoid things that appear unusual, or perhaps would alter their perception. Continual reinforcement of the change is the best counter to this type of resistance. When the benefits of change are continually reinforced, it becomes difficult for resisting groups to band together to form resistance. Create slogans or market the change internally so people are aware of the impact and importance of the change.

In the end, these are just some good examples, and some suggestions to overcome some of the resistance that a manager might encounter. The reality is that the manager must modify his or her style to accommodate the situation. There is no one right form of leadership, just as there is no one right organization. The more techniques a person can learn, the greater the flexibility in meeting the ultimate goal of leading change.

> Practical Tip: Make sure to talk to the team about change. Ask them how it makes them feel and listen to their concerns. The more that can be done to make change easier, the faster the change will take place.

Ethics and the Virtual Project Manager

Ethics and leadership share the perception of *I know it when I see it* but academics and project managers have a difficult time succinctly defining both. So, how does a project manager establish ethics on a virtual project when it cannot be defined? The United States has passed some legislation to govern some behavior as it tries to offer some level of ethics standards for its citizens. The PM's company may have a vision or mission statement, processes/procedures, and a culture that may provide guidance. However, how does the PM provide the leadership and the ethical foundation for a project team that may include many cultures from around the world? This difficult situation requires exploring and understanding some the issues surrounding ethics and the culture of the project.

The virtual project depends on technology to communicate, is separated geographically, and the project team rarely meets face to face. The project team may be separated by a few miles or by continents. Leadership develops and provides the vision and strategy for the project while motivating the project team to embrace and implement them. Normative ethics are the practical aspects of ethics that regulate a person's moral standard. It helps the individual decide on right and wrong behavior. Applied ethics examine controversial issues and may result in establishment of legislation or rules.

The Project Management Institute™ (PMI) is the major professional society for project managers. A conclusion could be drawn that PMI would drive the ethical expectations of the project manager. There is a noticeable lack in defining ethics or even addressing ethics and a project in the various standards and the *PMBOK® Guide* published by PMI. Perhaps

PMI's leadership believes that the Code of Professional Conduct (referred to as the Code) is sufficient. The Code mentions ethics once, in regard to reporting conduct violations. More importantly, the Code provides a prescriptive view of professional conduct (rules based) that is separated into two major categories, responsibilities to the profession and responsibilities to customers and the public.

The United States and several other Western countries have established legislation to drive ethical behavior. Most notably, in the United States the project manager needs to be aware and understand the Racketeer Influenced and Corrupt Organizations Act (RICO), Sarbanes-Oxley (SOX), and the Foreign Corrupt Practices Act (FCPA). Each of these pieces of legislation was enacted in response to unethical behavior and the lack of self-regulation within some companies based in the United States. The project manager needs to understand how the legislation may have to be enacted on the project. Things to ask include but are not limited to the following:

Are legislative requirements built into the project management methodology that I am using?

Do I need to create processes/procedures to enforce the legislation?

Do I need to provide training for the project team members on ethics as it pertains to this project?

Which legislation is applicable to the project (e.g., is the project global or domestic)?

Will cultural norms be an issue in other countries (e.g., is bribery commonly used to conduct business)?

Does the company I work for have a code of conduct/ethics that all employees must adhere to? Are the vendors/subcontractors required to adhere to the company's code of conduct?

Does the company have a culture that will help or impede ethical behavior on the project?

The project manager needs to understand his or her own ethical behavior and how he or she will implement ethics on the project. Will ethics be implemented via processes and procedures? Will ethics mean obeying the law and corporate policy? Will it be a combination of rules with ethical leadership? Each of these items will establish a tone and underlying current on the project. Total rules-based ethics has its pros and cons, as does the leadership approach.

Rules-based ethics may drive the project team members to view ethics as white or black. There are gray areas regarding ethics that rules and legislation cannot anticipate. In a rules-based ethics environment, the greater the intrinsic value, the greater temptation there is to bend the rules. U.S. financial practices are heavily regulated; however, there is a vast amount of rule-bending and unethical behavior. Enron's and Worldcom's issues stemmed from financial dealings and making the companies appear healthier than they were. Some of this rule-bending resulted in legislation.

Ethics via leadership is the most difficult to implement on the virtual project. The project manager must understand his or her own leadership style and ethical behavior, and constantly uphold ethics. This form of ethics is centered on trust. In fact, trust is the pivotal factor for a successful virtual project. The ethical leader has to inspire trust (Business Ethics, 2004; Ethics & Policy Integration Center, 2007). The project team has to know the project manager can be trusted to set the example of ethical decision making, inspire others to look for the ethical solution, and not tolerate those who violate the trust of ethical behavior and decision making.

What is the best means of establishing ethics on the virtual project? The project manager has to balance the law, the organization/company culture, the countries involved, and his or her own ethical behavior. This is never an easy solution. In most cases, the project manager needs to implement a hybrid solution. Uniting various cultures, countries, and organizations into a new unit requires an ethical foundation. The project manager should take advantage of any available online ethics training and should make this a requirement. If none is available, the project manager needs to take the time to create a PowerPoint presentation that details the legislation applicable to the project.

> Practical Tip: Keep in mind that ethics are not absolute worldwide. What we consider ethical may not be ethical in another country. Become familiar with the ethics of any other cultures or nations that you are frequently in contact with. This will help you better respect and understand the ethical perspective of others.

Once the training is done, the project manager must implement and take on the role of the ethical leader. Some might believe that ethical leadership is gentle and forgiving. However, research demonstrates otherwise. The leader has to make sure that he or she does not create resentment for being tough and autocratic. Altering leadership styles can assist the project manager to create the trust and perception that the ethical improprieties are addressed appropriately.

Five different leadership styles will assist the project manager in handling various ethical situations. The suggested styles adapted from EPIC (2007), Business Ethics (2004), and Block (1993) are shown in the below table.

Leadership Style	Description
Setting the example	The least intrusive. Lead by example. Even in the virtual world it can be done. Set tone for meetings. Set tone for e-mails. Set tone for processes/procedures. React appropriately during ethical violations or ethical gray areas.
Guidance/Stewardship	The teaching leader. Analyze the ethical situations with team members. Occasionally provide case studies with possible solutions. Be there to help; do not make the decision. Offer constructive criticism for ethical mishaps. Update plans/processes/procedures as appropriate.
Convincing	The persuasive leader. Rally the team to a common goal. Appeal to reason. Use critical thinking to show the correct solution. Have other team leaders demonstrate their resolve to support your position.
Incentives	The rewards/punishments leader. Do not use often. Can be offensive to some. Leader truly needs to evaluate the ethical "message" being introduced. Commonly used for black-and-white situations.

Leadership Style	Description
Authoritarian	The most intrusive. Should be used sparingly. Forces someone to do something he or she is not committed to do. Use for actual or potential ethical breeches (legislative, corporate, or project). Be decisive and direct. Maintain confidentiality but ensure the message to the team is clear: Ethical violations are not tolerated. Have lessons learned, if needed. Update plans/processes/procedures appropriately.

Leadership is the pivotal foundation of the ethical character of the project. Rules, legislation, processes, procedures, and management plans may create a very black-and-white perception of ethics. The U.S. financial industry is a prime example of how legislation is only a stopgap effort to help regulate ethics. When individuals perceive ethics as not breaking the law, then legislation is in a constant catch-up cycle. Leadership needs to establish the ethical tone or soft criteria for whatever falls into the gray areas.

> Practical Tip: Always consider how you would feel if your actions were the headlines of tomorrow's newspaper. If you feel that the story would be harmful to you, your team, your reputation, or the project, then you should consider carefully the ethical nature of your actions.

COMPLEXITY IN THE VIRTUAL PROJECT

A common theme throughout virtual project research is that a virtual team is not simply a group of individuals. A common goal, purpose, and interdependence are characteristic of a team (Axelrod, 2002; Frame, 1995; Katzenbach & Smith, 1994; Kerzner, 1998). A successful team relies on each member to achieve the common goal. All of these elements are necessary for a successful virtual project team. Furthermore, identifying and harnessing these kinds of relationships are critical for complexity. Complexity in the virtual project is about the complex interpersonal relationships that make up the virtual project. Understanding these social networks and focusing upon them the project is one of the important manners in which complexity can make a virtual project more successful.

Within business, team members are normally assigned by management. Axelrod (2002) points out that within business, it is perceived that the team members did not have a choice. However, Axelrod states that team members have three choices:

1. Participate willingly,
2. Sabotage the team, or
3. Leave while staying in place.

When team members choose to belong to the team, they normally perform the work with care. Understanding this feeling is part of the complex arrangements of projects. The project manager must understand that all the members of the project team or all of the stakeholders are not always fully behind the project. Katzenbach and Smith (1993) states that a project often fails because; the team is only a working group instead of being a fully integrated team. The primary difference between a working group and a team is that a working group does not have a common purpose, nor is it self-defining. Furthermore, the senior ranking individual, not necessarily the true leader of the team, likely leads a working group. A working group has self-interests and has to report progress and issues to a supervisor.

Research by Sethi, Smith, and Park (2001) contradicts the findings of Katzenbach and Smith. The research indicates a positive correlation between senior management oversight and cross-functional teams that are successful in new product development (Sethi et al., 2001). Holland, Gaston, and Gomes's (2000) qualitative research supports Sethi et al. (2001). Both studies state that senior management oversight is productive for cross-functional teams, as long as risk-taking is encouraged: "virtual teams entail much more than technology and computers" (Duarte & Snyder, 2006, p. 9). They include seven factors for success, and technology is only one of the factors. The other six are as follows:

Human resource policies
Training and on-the-job education and development
Standard organizational and team processes
Organizational culture

Leadership support of virtual teams
Team-leader and team-member competencies (Duarte & Snyder, 2006, pp. 12–13).

Maznevski and Chudoba (2000) suggest that leaders of effective virtual teams require face-to-face meetings initially to build camaraderie among members of the team. Jarvenpaa and Leidner (1999) also suggest that face-to-face meetings should occur periodically for effective virtual team leadership. Hinds and Bailey's (2000) studies support the finding that face-to-face meetings enhance the success of virtual teams. Trust is essential to the virtual environment's success, and the face-to-face meetings increase the trust among the participants (Maznevski & Chudoba, 2000; Jarvenpaa & Leidner, 1999)

Training and learning are major factors in the virtual environment's success (Townsend & DeMarie, 1998; Duarte & Snyder, 2006). A virtual team appears to be more successful when training is conducted on communication skills and communication technology (Townsend & DeMarie, 1998). Townsend & DeMarie's (1998) studies indicate that technology training should occur more often for virtual teams than for traditional teams since technology is the mainstay for communication and is evolving at a fast pace. Duarte and Snyder (2006) find that a successful virtual project team distributes "learnings within the team and beyond to the wider organization" (p. 127).

As a result of their study, Roberts, Kossek, and Ozeki (1998) find that executives dealing with virtual projects have three common issues: ensuring the correct skills are in the correct region/area when needed (¶13); disseminating innovative and "state of the art knowledge and practices" (¶14); and identifying the talent throughout the organization (¶15). English is the business language for all the companies within the study. However, this did hinder the virtual organization because of the different English grammar, English not being a native language, and the nuances of the various English versions.

The ability for the virtual project manager to encourage and accept change within the team on complex projects would be seen as encouraging creativity and growth (Singh & Singh, 2002). By contrast, the need to control and maintain onerous processes and procedures encourages the *as is on a project,* which research suggest does not work on complex projects (Cooke-Davies, Cicmil, Crawford, & Richardson, 2007; Overman & Loraine, 1994; Singh & Singh, 2002). Singh and Singh suggest the project manager should focus the team leaders on "dynamic instability" (2002, p. 31). Essentially, the project manager should encourage the teams to look at other alternatives. Remember, if your team project does not, there will be other companies that will encourage their projects to do so.

> Practical Tip: When considering the power of complexity theory, keep in mind the metaphor of a swarm of bees or a swarm of ants. Each individual is motivated toward a single goal, but each ant or bee will each take its own path toward that goal. The organization seems outwardly chaotic, but the results can be devastating as a swarm of bees or ants routinely take down opponents many times larger than any individual in the swarm.

THE CHAMPIONS OF COMPLEXITY

Successful virtual projects are seen in many industries. More companies have established processes and procedures for the virtual project and provide appropriate training. Research demonstrates that a centralized Project Management Organization (PMO) is more beneficial for the virtual project manager, from which can be extrapolated that virtual projects would see benefit as well. Two examples of successful virtual projects are presented below.

Case Example: Software Consulting

A large firm was developing a software upgrade for a bank's ATM. The project was an international firm and used "follow-the-sun" software development. The project was initially a disaster. Each of the development areas was using its own methods to compile the software. By the time the software returned to the United States, the software code seemed to be completely different. However, the lab tests indicated the software was working properly.

The software was implemented into the ATMs and it did not work correctly. It quickly developed into blame. The company that developed the software accused the bank of making changes to the hardware and the bank accused the company of faulty coding.

A recovery project manager (RPM) was sent to review the project. The recovery project manager was adept at virtual projects and had worked on complex IT issues in the past. The RPM quickly understood that the fault was with the company, not the bank. A communications plan was put into place with processes and procedures to clearly explain the handoff procedure for code and the compiling process. Each lab was equipped with the exact same software and hardware.

Within six months after implementing the new processes and communications plan, the project was delivered. The project manager instituted a campaign to ensure that the bank was happy and would continue to use the company for further software development. The recovery project manager was not colocated with the bank. In fact, the recovery project manager never met the bank's sponsor. The recovery project manager instituted a communications plan that accounted for the complexity of the virtual project and ensured that all levels were accounted for in the plan, thus leading to the project's success. This example highlights the importance of a communication plan for a successful project. Teams that are not communicating are wasting resources and creating confusion in the project.

> Practical Tip: Many organizations spend a lot of energy affixing blame to individuals. It is more productive to offer solutions than blame. In the end, blame is rarely the fault of a single individual and so it becomes difficult to find the one person to blame. Instead, spend more time with the solution because that will ultimately be more responsible than who was to blame.

Case Example: Online University

As an online university implies, it is virtual. Many online universities create their curriculum with adjunct faculty members who are not present at the university. The development is done virtually. It is very well organized. For a six-week course, a university will normally allocate twelve weeks. The course project manager has a discussion with the course developer before the work starts. Via this discussion, the course project manager understands how the course developer works. This then allows the project manager to clearly lay out a project plan and schedule for the course development. This method works because there are numerous conference calls and e-mails to ensure the timetables are met. Should the adjunct faculty member fall behind, the project manager increases the number of phone calls and will provide tips on getting the work done. There are milestones built into the process where the project manager can request that the adjunct faculty member be replaced.

> Practical Tip: Online universities are more common than ever before and much can be learned by this new method of education. Keep in mind that the example above can be applied to students and other projects. Often the difference between the good and the great are those who do not abdicate responsibility but help others achieve greatness.

THE MUTTS OF COMPLEXITY

Not all projects are successful, nor do they always bring value. Some projects bring no value to the organization and often cost the organization more than it is worth. As with the champions, it is important to take communication into account. With many failed projects there are communication issues. The less structured the communication, the more likely the project will fail. Understanding and taking into account the basic elements of a project are critical for success. The PM must make sure to consider these factors. We will review some projects that did not take these important factors into consideration and we will find how those projects did.

Case Example: Introduction of a Project Management Organization (PMO)

A global financial organization had decided to implement a corporate IT project management office. The CIO had no sense of the types of projects, nor how many were being done in the various business units. There was a sense that there were many duplicative projects. The CIO started to implement this new organizational shift without the buy-in of senior management or his peers in the business units.

A project manager started to implement the PMO for a virtual environment. Roadblocks soon became the standard process on the project. This project involved senior managers and project managers from several countries.

The entire project failed because the project manager forgot to take into account that an organizational shift is a project in and of itself. Since senior management in the business units had not accepted this shift, the project managers with the information were not allowed to share the information.

Case Example: Network Infrastructure Paradigm Shift

A hospital network IT infrastructure was a series of spiderwebs. Each system had to be changed whenever a change to the infrastructure was needed. Many of the systems were outdated and had to be hard coded. The system had become so complex that there was no complete description of the systems and how they interconnected.

It was decided to investigate whether implementing an entirely new IT infrastructure was feasible and cost effective. The CIO of the hospital network's IT department selected individuals who were in favor of the project. The CIO also ensured that there were some naysayers to achieve balance.

The communications plan was very detailed since many of the potential vendors were located in other areas of the country. The hospital network's IT facility was located in four different states. There were representatives from each facility. Each representative had to recruit individuals to help develop use cases for the different scenarios. Templates were developed to have the use cases appropriate for a due diligence and not an implementation.

A collaboration Web site was implemented on the project. This allowed the team to be able to access the latest documentation, including risks, issues, schedules, and minutes. This decreased the amount of e-mails and helped to ensure that all parties were using the latest documentation. The technology was not the latest but everyone was able to access it.

CHAMPIONS VERSUS MUTTS: KNOWING THE DIFFERENCE

The common factor in the success and failures of the above projects is trust. The projects that lacked trust ended in failure, or at best in marginal results. The projects that had a high degree of trust resulted in success or at least in a positively perceived project. There were certainly other factors but in the end all roads lead back to trust.

In any failed project, there is ample opportunity to review the negative results and to consider alternative paths. In these discussions, the following should at least be considered: trust, communication, technology, conflict, team formation, allowing some chaos, dual project plans, clear processes/procedures, and clear strategies and goals, as well as looking at a centralized or decentralized project organization.

PROJECT SUCCESS OR FAILURE: COMPLEXITY—LOVE IT OR LEAVE IT

Project success depends upon the team leader's ability to harness the available resources, the ability of the team to communicate to the organization, and the ability of the stakeholders to remain focused toward the common goal. All of these elements are necessary for success. A project that is lacking in any of these elements will be challenged to attain success. In order to better understand project success, the diagram in Figure 4.2 has been developed to show the interdependencies between the project, the team, the leader, and the stakeholders. In short, the graphic explains how the leader must drive the project, stakeholders must support the project, and team members must remain enthusiastic and energetic toward the project.

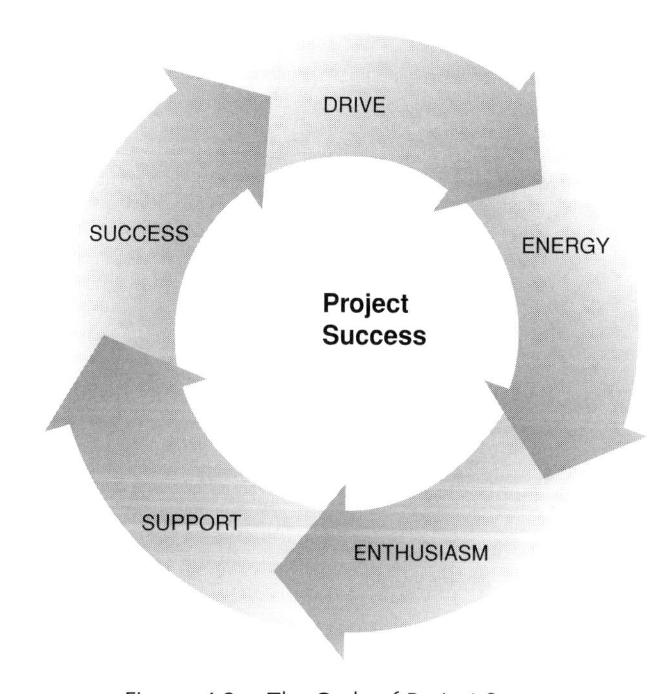

Figure 4.2 The Cycle of Project Success

> Practical Tip: Leaders need to operate as a focal point of the project. They must project the projects' goals by communicating the requirements of the team to others while meeting the challenges of the project.

Drive

A leader must drive the project, the team, and the stakeholders toward success. Everyone who comes in contact with the leader can become an enthusiastic supporter of the project. Everyone the leader communicates with can become an informed messenger of the project. Research shows that leadership is the defining and critical element in successful projects.

Leadership is critical for the success of virtual teams. However, task leadership is not sufficient for success in the virtual environment. In addition to a virtual team needing leadership, several studies found that the successful virtual PM is competent and adept at the following: developing and transitioning team members, developing and adapting organizational processes to meet the team's needs, allowing leadership to transition, and ensuring the team received necessary training for virtual communications and technology and skill sets.

The PM should establish metrics that measure team performance. These metrics should institute high expectations to encourage the extra effort required to overcome the communication hurdles. As the team becomes familiar with the team expectations and the performance metrics, the amount of undesirable conflict should be reduced. Ultimately, a leader must not only steer the direction of the project but must drive the subordinates toward the final project goal.

> Practical Tip: Once you create a metric that measures team performance, you must refer to it, tie it to performance reviews, and keep it visible in order for the team to rally around the requirement.

Energy

The team must have a sufficient level of positive energy to bring everyone together toward the goal. Leaders must find a way to give purpose to others to expend and focus energy to achieve targeted goals (Godin, 2008). Leaders can set high goals to focus others, but teams must then meet the challenge. Sometimes setting too low of goal or setting an unclear goal can demotivate a team.

Setting precise goals is important. Too often a goal is created by mutual acceptance, rather than trying to create a stretch goal. For example, in one organization, by setting profit targets rather than sales or productivity goals, one team was able to manage a string of 40 consecutive profitable quarters. This reframed the success goal (sales) to a high-energy goal (profit). A high-energy goal can often motivate team members to achieve greater success.

> Practical Tip: The leader can set the tenor and tone of the energy for the project. By communicating in bursts of energy and keeping a consistent message, the leader can help everyone move toward the same result. You can use the same strategy as is used in marketing and commercials. Offer short vignettes that explain the goals of the project while making sure to communicate the values of the organization at the same time.

Enthusiasm

Enthusiasm is transformational and revolves around communication. One must be able to transform people from team members into enthusiastic project members. Those involved must want to be part of the process and must want to tell others about their transformational experience. Enthusiasm is linked to the transformational nature of individuals. Leaders need to be transformational, to understand the needs of the team members, and to pass on knowledge so a project can be successful.

Enthusiasm is contagious. Being around enthusiastic people creates more enthusiastic people. Attending any sporting event will show how contagious enthusiasm is. At one point, some colleagues had decided to attend a bullfight to see what it was all about. Several of them did not feel that it was a humane sport, but they were curious and good seats were available for the event. After being in the stadium for a while and feeling the energy as the matador thrilled the crowd, entire groups of noncommittal participants were cheering along with the crowd. The enthusiasm of the event had brought forth support for the event.

Enthusiasm is communication. There must be an understanding of the critical differences between teams and virtual teams. Traditional teams communicate directly through face-to-face contact. This direct communication includes body language, tone, and other visual cues, which are important elements of communication. Virtual teams regularly contend with the lack of direct contact and rely upon indirect communication, in the form of telephone calls, e-mails, faxes, and other technologically based methods (Duarte & Snyder, 2006).

The virtual environment requires that communication be clearer and more concise. Technology cannot replace poor communication; in fact, effective communication is one of the most critical elements of a virtual team. Furthermore, the leaders of the organization must support the virtual team by providing the necessary technological resources for these types of projects. To overcome these challenges of the virtual environment, many studies and documented practical experiences emphasize the need for trust in the virtual environment.

> Practical Tip: Genuine enthusiasm is an effective way to motivate a team. If the goal is interesting enough and is supported by the organization, then it is possible to make it something to get excited about. Never force enthusiasm but allow it permeate the team.

Support

Support of a virtual team is essential. Two important elements of support are the ability to gain customer and stakeholder support for the project and the access to technology. Time and time again, successful virtual project managers gain support of customers and stakeholders; however, state-of-the-art technology is not necessary for success. Access to comparable electronic communication technology is important, but no research appeared to support the need to for state-of-the-art equipment. Virtual teams are more successful when they are linked to an extended network of experts (stakeholders) that support the group. The PM identified these stakeholders and gained their support in order to make the project successful. The creation of a stakeholder matrix outlining the level of participation, roles, and contact information is important to allow all team members to understand what resources are available to the team. Therefore, PMs should understand their own competencies and make sure that they are appropriate for their team. If they lack a critical element it then becomes important to ally themselves with another project team member who has those important elements.

Practical Tip: Create a newsletter about the project to generate support for it. The more publicity there is for the project, the more energy will be focused in that direction.

Thus, project success depends upon the skills and ability of the project manager. Hence, it is critically important to review, study, and hone these skills in order to better leverage them in difficult projects. Also, one needs to vary the application of these elements in order to be more effective. Simply trying the same failed application over and over again will not yield improvement. Learn to vary the approach, style, and communication in order to become more effective.

SUCCESSFUL VIRTUAL TEAM INTERACTION: ASSESSMENT, CONSEQUENCES, AND MANAGEMENT

A successful virtual team is usually composed of experts, with these individuals exhibiting three types of behavior: constructive, passive, or aggressive. Constructive is a style that places priority upon creating an overall positive atmosphere, with an emphasis upon trust and encouragement. Constructive behavior offers a balance between personal and group outcomes. Passive behavior is described as being less communicative, with an emphasis on the fulfillment of goals only. Passive behavior also places harmony above other objectives, which often leads to limited information sharing and limited ability of individuals to challenge the input of others. Aggressive behavior places the highest emphasis upon personal achievement and personal ambition. Aggressive behavior also stresses competition, interrupting others, and impatience.

These three types of virtual behavior can lead to mixed business results. Overall, in the virtual environment constructive behavior offers the most consistently outstanding performance. The passive and aggressive styles tend to have more varied results, with the aggressive style having more outstanding results than passive behavior. From a management standpoint of virtual teams, this is an excellent tool to determine if management intervention is necessary when monitoring virtual teams. If a manager observes either passive or aggressive behavior within a virtual group, the manager can then delve deeper into the details to make sure the team remains focused and on track.

The virtual team is considered a vehicle for modern business, with several key business justifications and shortcomings for such an arrangement. Business justifications include documented increased productivity of virtual teams, the ability to access global markets by crossing national boundaries, and the positive environmental impact of reduced automobile emissions. Business shortcomings include the high cost to set up and maintain individual employees in home offices, the loss of concentrated cost efficiencies, cultural challenges, empowerment challenges, and communication challenges, as well as workers feeling isolated from the organization.

The critical factors for management success are seen as effective and efficient delegation, results-oriented goals, scheduled virtual meetings, and a commitment of swift communications. Additional challenges include technical issues, which can serve to exacerbate communication; lack of clear role definition in the forming stage of the team; lack of a positive attitude in all communications; and the lack of less meaningful social interactions, which typically occur around the water cooler in traditional companies (Duarte & Snyder, 2006). In many cases, good management techniques translate to good management techniques in the virtual world. The difference is that the principles of good management must now be followed more closely than before, because the opportunity for communications errors or miscommunication is greatly increased in the virtual environment. Successful virtual managers must look forward to lead change, rather than looking backward and continually reacting to change.

HOW COMPLEXITY CAN ACHIEVE GREATER SUCCESS

The project manager cannot be in all places at all times. Traditional best practices have not been successful on large complex projects. Anecdotally this has been seen with the Standish chaos report and this has also been seen in academic research over the years. Complexity theory has provided some indications that complex projects do better when certain aspects of complexity are applied.

The *butterfly theory*, six degrees of separation, and the patterns within chaos will help the complex project survive, especially those in the virtual environment. These three major subtheories can help the project manager decide where to focus the traditional best practices and where to allow complexity theory to take place and just to monitor progress. The project manager has to understand that complexity theory is a part of nature and has to become comfortable with its effects.

Trust Experiential Review

Jonathan McDonald's career in the electronics business was propelled to the next level when he became the youngest general manager for Custom Electronics in 1983. During this time, the smaller electronics firms were being purchased by the larger firms. He had been successful for years at Custom Electronics, but Jonathan began to consider other opportunities. In 1989 he joined a start-up electronics company, Alpha & Omega Electronics. Alpha & Omega Electronics was to become the leader of the industry and take high honors in areas of electronics entertainment and customer service. After the successful launching of fourteen new stores, Jonathan then moved forward to become the managing director of World's Best Electronics in 1995.

Jonathan McDonald is a transformational leader who puts a high emphasis on productivity while still listening to his subordinates. He displays the self-confidence and courage of his convictions and vision. He also has traits of a change agent, and will accept whatever task is necessary for the goals of the organization. Since he has been involved with new companies, he has shown himself to be a leader who not only can motivate by example but also can change, as the circumstances require.

As start-up companies are very demanding, Jonathan ranks productivity as critical to the success of any organization, and considers people issues a low priority. Jonathan may not always know the details of the organization, but he will monitor the situation from afar, as long as the internal and external reports are positive. He feels that his team must outshine all others in order to continue his string of successes. He feels that it is important to be surrounded with top performers. His lesser concern for people becomes more apparent in times of crisis; he is more often likely in times of trouble to avoid contact with his people than when operations are running smoothly.

Jonathan has directive behavior and thrives on telling people who, what, when, and where to perform actions. He leaves the details to the individual, and he is comfortable giving high-level directions with inflexible deadlines. He demands immediate results when he requests a task to be completed and excuses are unacceptable. If project delays are possible, it is best to advise him immediately rather than to conceal this type of information. Feedback is critical to Jonathan; if he does not hear back he will assume that all is well.

He is a focused, charismatic leader. He will explain his vision of what the organization will look like so people understand what the future will look like. Jonathan is best described as a leader and not a manager. He has honed his skills in influencing groups of people, rather than just individuals. Over time, he has crafted an image and an impression that he gives to all who come in contact with him. This allows for him to create a perception of confidence. In the chaotic environment of a new company it is particularly important for others to be able to anchor themselves.

He is a formal leader who finds this formality necessary to interact with people of all levels. This style creates an aura that helps his position in negotiations as well as dealing with his peers. Creating a layer between himself and those with whom he normally comes in contact allows him to be efficient and businesslike at all times. Part of this is due to his Norwegian heritage' Norwegians are often perceived as distant and cold. This is not truly his personality; the first impression given is one of a distant and authoritarian leader.

Jonathan is largely driven by legitimate power. He feels there is a level of title power given by an organization. To a large degree this is true, but this is a weakness of his. Often a lower-ranking individual will defy him, and he is often baffled at how to deal with such a circumstance. These people can frustrate him because he cannot understand why they do not respond, despite his repeated requests. This weakness comes from his lack of exercising other forms of power. If only legitimate power is used in all circumstances, there will be times that goals are not met and targets are not accomplished.

Jonathan has an accommodating style. He does not feel that anyone has ever achieved more by having an adversarial relationship. Jonathan understands that there is a time and place for change and conflict; he does not believe that it answers the questions that caused the problem. Problem solving is more important to him than individual's hidden agendas. Often he becomes the mediator in a conflict and accommodates his goals to meet the goals of others.

Brainstorming is his preferred method of group decision making. He believes that more is better; however, he often does temper this by excluding some and including others. He does not utilize all levels of the organization and he relies on his direct reports bringing all available information to these sessions.

Thus, Jonathan is a bold and dynamic leader; however, his style is not for everyone. Consider the information from the chapter, and write an essay that encompasses the central question of whether or not you would trust this kind of leader.

Why would you trust this person as a virtual leader?
What do you trust about this person as a virtual leader?
What do you not trust about this person as a virtual leader?
Do you feel that this is the kind of person you would want to work for? Why or why not?
Do you feel that this is the kind of person you would want as part of your organization? Why or why not?
Do you feel that this is the kind of person who would do well in your current organization? Why or why not?
How is this person like you? How is he or she not like you?

CHAPTER SUMMARY

Trust must be part of the foundation of a virtual organization. Trust demands boundaries and learning, and it requires that leaders and followers are bound together in a manner where they both benefit from the relationship. Trust must be visible at all levels of a virtual organization in order to be successful. Successful virtual teams require that camaraderie be built among the members of a team. The leader of the virtual team must also conduct himself or herself as a strategist and create an environment conducive to trust. Once that environment is achieved, then organizational change will happen smoothly and effectively.

Ethics are important for the virtual team because they define the organization at all levels. An ethical framework is important to make sure that the organization is not only compliant with the law, but the team is compliant with the goals of the organization. Maintaining a trusting and ethical organization will help steer project members toward success when there are no clear guidelines to the situations that they encounter.

REFERENCES

Axelrod, R. (2002). Making teams work. *The Journal for Quality and Participation*, *25*(1), 10. Retrieved May 17, 2002, from Proquest.

Block, P. (1993). *Stewardship: Choosing service over self-interest*. San Francisco: Berrett-Koehler.

Business Ethics: A manual for managing a responsible business enterprise in emerging market economies (2004). Washington, D.C.: U.S. Department of Commerce: International Trade Administration.

Cooke-Davies, T., Cicmil, S., Crawford, L. & Richardson, K. (2007). We're not in Kansas anymore, Toto: Mapping the strange landscape of complexity theory, and its relationship to project management. *Project Management Journal*, *38*(2), 50–61.

Duarte, D., & Snyder, N. (2006). *Mastering virtual teams: Strategies, tools, and techniques that succeed* (3rd ed.). San Francisco: Jossey-Bass.

Ethics & Policy Integration Centre (EPIC). (2007). *The role of leadership in organizational integrity, and five modes of ethical leadership*. Website http://www.ethicaledge.com/quest_4.html

Frame, J. (1995). *Managing projects in organizations*. San Francisco: Jossey-Bass.

Hinds, P., & Bailey, D. (2000). Virtual teams: Anticipating the impact of virtuality on team process and performance. *Academy of Management Proceedings*, 1.

Holland, S., Gaston, K., & Gomex, J. (2000). Critical success factors for cross-functional teamwork in new product development. *International Journal of Management Reviews*, *2*(3), 231.

Jarvenpaa, S., & Leidner, D. (1999). Communication and trust in global virtual teams. *Organization Science*, *10*(6), 791.

Katzenbach, J., & Smith, D. (1994). Teams at the top. *McKinsey Quarterly*, *1*(71), 9.

Kerzner, H. (1998). *In search of excellence in project management*. New York: Van Nostrand Reinhold.

Krajewski, L. & Ritzman, L., (1996). *Operations management, strategy and analysis* (4th ed.). Reading, MA: Addison-Wesley Publishing Company.

Maznevski, M., & Chudoba, K. (2000). Bridging space over time: Global virtual team dynamics and effectiveness. *Organization Science: A Journal of the Institute of Management Sciences*, *11*(5), 473.

O'Toole, J. (1996). *Leading change: The argument for value based leadership*. New York: Jossey-Bass.

Overman, E. & D. Loraine (1994). Information for control: Another management proverb. *Public Administration Review, 54*(2), 193–196.

Project Management Institute (Ed.). (2008). *A Guide to the Project Management Body of Knowledge—Fourth Edition*. Newtown Square, PA: PMI.

Project Management Institute (Ed.). (2009). *Practice standard project risk management*. Newtown Square, PA: Project Management Institute.

Roberts, K., Kossek, E., & Ozeki, C. (1998). Managing the global workforce: Challenges and strategies. *Academy of Management Executives, 12*(4), 93.

Sethi, R., Smith, D., & Park, C. (2001). Cross-functional product development teams, creativity, and the innovativeness of new consumer products. *Journal of Marketing Research*, *38*(1), 73.

Singh, H., & A. Singh, A. (2002). Principles of complexity and chaos theory in project execution: A new approach to management. *Cost Engineering*, *44*(12), 23–33.

Standish Chaos Report. (2009). Standish Group.

Townsend, A., & S. Demarie (1998). Keys to effective virtual global teams. *Academy of Management Executive, 12*(3), 17.

Part II

How to Deploy Complexity Theory

Part II provides the practical application regarding complexity theory. This section will review the effective use of complexity within virtual projects. Complexity is not appropriate for all aspects of a virtual project; however, certain key areas require more dynamic management. By its very nature, the virtual project needs to have some areas that have rigid processes and procedures. By contrast, there are other aspects of virtual projects that lend themselves readily to complexity and are discussed in this section. Understanding this critical differentiation and application can only improve the project manager's success rate as well as improve the organizational agility with regard to handling the difficult and complex changing requirements of difficult projects.

Chapter 5

Successful Project Management Strategies of Complexity

Figure 5.1 A sunset is an expression of complexity in the natural world. No two sunsets are exactly alike, yet everyone recognizes a sunset when they see one. Note the depth and variety of color that permeate this sunset. Complexity can offer the same type of depth and variety in project management.

PLANNING FOR COMPLEXITY IN A PROJECT

Planning for complexity in a project is no different than how a plan is created for any other function or process in a project. In all projects, managers must plan for the uncertainty of the future (Titcomb, 1998). This involves budget forecasting, labor planning, and organizational capital acquisitions, such as in the normal management planning process (Krajewski & Ritzman, 1996). Managers must forecast their future needs by offering the organization detailed information regarding the financial situation of their department or team. Labor planning is accomplished by alerting the organization to any changes in their productivity or production needs. Organizational capital acquisitions are any new technology that will be required to be more efficient or to perform future forecast tasks. Managers must plan contingencies and have action plans to address issues that arise.

The importance of planning is to allow for the future actions to be actively considered prior to making any form of financial commitment. Organizations have many specific planning types and functions; the objectives of the planning aspect of an organization remain the same. Planning is the thinking that management must do prior to taking action. However, planning for complexity is often difficult. Since it becomes a less directed matter and it is often related to change, it seems to be a more ethereal process. Managers must continually plan for change in order to be prepared for the forces of uncertainties in the future (O'Toole, 1996).

To understand how complexity is to be planned for, one must first understand the management planning process, which is well documented in the *PMBOK*® (PMI, 2008). It is important to review how this process can be accomplished, even in an organization that might not embrace complexity as a sound business theory. Management must weigh the different alternatives before moving forward with a plan. In essence, the management planning process is the manner in which leaders enact change within an organization. In the management planning process, objectives may change at different management levels; the process remains the same.

The management planning process has five distinct phases. Each phase requires management intervention, and each step requires action and input, which will ultimately build toward the final decision and plan (Krajewski & Ritzman, 1996). To understand how this process works with complexity, there will be a definition of each phase in the management planning process, followed by a review of how this fits into complexity.

1. Establish Objectives. The establishment of high-level goals is the purpose of this phase of the management planning process (Rigby, 2009).
2. Analyze Situation. This is the phase where all information regarding the current situation is reviewed. Options are categorized into areas

of benefits and losses when compared to the present and future situation (Rigby, 2009).

3. Determine Alternative Courses of Action. Management must offer different ideas to achieve the goal. Regardless of the objective, there will always be different methods to achieve similar results (Godin, 2008).
4. Evaluate Alternatives. The manager must establish objective criteria to compare the different options available. These criteria will determine the relative worth of each alternative (Godin, 2008).
5. Choose and Implement Plan. From the various alternatives, the manager must weigh all the organizational factors carefully and then the best alternative, based on the objective criteria, is selected and implemented (Brown & Eisenhardt, 1997).

This linear and contingency-based process does make one wonder how can it be used to implement complexity effectively. The soundness of the above process is what has made it so effective for so long. However, there have been so many changes in projects and project management that one cannot continue to use this type of model. In the past, projects were fairly static and milestones were cast in stone. Furthermore, there was a single dedicated project manager for a single project. These kinds of assumptions are no longer correct, so there must be a more dynamic process to manage projects. Even beyond those limitations, the fundamental flaw in this process is that it assumes all the alternatives can be evaluated at a single point in time and there will be only one plan that is the correct one.

So, given the above criticisms of the contingency-based planning process, where does it leave the project manager? Does this mean one should never develop a project plan? Does it mean that planning is actually a waste of time? Does it mean that everything about project management organization that has been gruelingly researched, implemented, and codified to assist project managers should be discarded? As terrible as it sounds, the answer may be shocking to some because the answer to all of the above is actually worse than one could believe. The answer is actually *maybe*.

Maybe. That is about as vague as it gets to project managers. The problem with *maybe* as the right answer is that it is so unsatisfying. It makes for a difficult way to manage a project if there is no one correct path. Yet, in the United States and in many of the capitalistic constitution-driven republics of the world, one finds that *maybe* is actually the answer.

To illustrate this point, let us consider two hypothetical governments and see which nation we would prefer to live in. Let's call government one Project Manager Nation (PMN). This nation was set up by a project manager who wanted to create a country where rules were absolute and there was no room for any shades of gray. The head of state of PMN is elected by the constituency, based upon the current needs of the nation. Hypothetical government two is much more in line with the constitutional government of the United States. The major difference is in the electoral process for the head of government, which will be discussed later. The second nation is the Country of Project Management (COPM).

PMN is a directed dictatorship where the head of state is charged with addressing the current need of the nation. The head of state is charged to address a given issue, as well as an approved budget and timeline. If the head of state were to fail to meet the required milestones or to overspend on the plan, he or she could expect to be replaced by the next suitable candidate willing to take the task. If the head of state felt that the best way to address a current issue (let's say socialized medicine) is to levy what he or she believes is the appropriate tax on appropriate groups, the constituency would then be forced to work under the decision and the decision could not be undone because it would undermine the initial objective (project) that was approved.

To better understand how this works in practice, following are a couple of examples to give a better feeling for what this kind of nation would be like. For example, if PMN was having economic issues, the nation might elect their best professor in economics. If PMN had a trade issue, the nation might then move to replace their current leader with their best expert in trade. The head of state would then be charged with completing the current goals of the nation in a structured and linear manner that could be explained to the constituency with charts and Gantt charts. The head of PMN would be empowered to take all of the necessary decisions to complete the goals of the nation, regardless of the social consequences, as long as the goals were completed according to the preestablished budget and timeline.

As previously mentioned, the nation of COPM would be very similar to the United States with all of the dynamic elements of the Constitution. However, regarding the election process of the head of state, instead of the existing party process the only factor considered for the position would be that the elected individual must be an expert in the platform of the party. Hence, the person elected would have to be an expert in the problems identified by the party. In theory, then, the top officials would be equipped to deal with the challenges of the constituency. It is true that they might be ill equipped to deal with navigating government politics, but since their expertise is so important the expectation is the individual would be able to manage a way to be successful.

For those who have not already seen the parallel, PMN is an analogy to the current view of project management, while COPM is more of a view of the application of complexity to project management. For some people, living in PMN is a very orderly manner in which national issues are quickly addressed and completed. It may be comforting to some, but many people despise dictatorships and feel that any individual who is part of governance should have some representation in the process. However, this kind of representation is considered a hindrance in project management where information is only pushed out rather than being pushed out as well as being pulled in.

So, how does this get us any closer to understanding how to plan for complexity? To understand this, one needs to take a closer look at the Constitution of the United States (or the constitution of most nations, for that matter). This document was designed to fulfill the needs of a changing nation that would remain timeless regardless of technology, population, environment, or any other factors. This document has remained through wars, rebellion, national crisis, natural disaster, depressions, and many other national and international situations that are too numerous to count. This type of document is a dynamic example of how a project plan can be structured to better achieve the goals of the project. Instead of limiting oneself just to milestones, tasks, resources, reports, and other linear functions, why not also include a dynamic document that will embody the vision of the team and stakeholders while offering a guiding vision toward the completion of the project? In a way, the PMI code of ethics is an attempt to fill this void, as well as an attempt to create a sense of a project charter in the *PMBOK®* (Fourth Edition). However, there must be a little more formality to this concept.

The following list shows how one can successfully apply complexity to a linear project.

Establish Objectives. The establishment of high-level goals is the purpose of this phase of the management planning process. This step essentially remains the same, because one needs to understand the objectives of any project as a first step.

Analyze the Situation Quickly. The project manager must review the available material quickly, and no in-depth analysis is necessary. The project manager should review the available information as quickly as possible in order to develop a flexible plan.

Develop a flexible plan. Create a project plan that contains the required milestones but offers only guidelines for achieving the required tasks. Spending too much time detailing all of the steps of the process is often a waste of time because these will change as soon as the project starts.

To some, this linear plan may still not cover how complexity can be successful; to further explain there is another analogy that might assist. Complexity is sometimes described as a flock of birds (Hass, 2009). One cannot understand the movement of the entire flock by examining the pattern and motions of a single bird. The only way to understand the rules of the flock of birds system is to examine the entire flock. For example, some rules of the flock could be to maintain the same vector as the next bird, land when the next bird lands, or maintain a safe flying distance between other birds. Thus, the developed plan would be based upon the entire system and not built up by the required tasks. Another important element to consider is that there is not always a linear change in project dependencies. One needs to create a high-level plan to keep the project guided in the right direction. This plan should be the one that guides the project to success, not necessarily the one that is task dependent. The plan should also include a process to handle the human interactions and the flow of communication to the team and stakeholders, as well as from the team and stakeholders.

Thus, one needs to create a project listing that outlines all of the required elements but keep it open as to avoid too many encumbrances of a traditional Gantt chart. Consider this plan to be the outline so it can address many of the social elements and interactions that one can expect. Think of this as the project outline that is designed to cover the different elements of the project and understand that these elements exist within a complex system with different interactions. This document can be used to supplement all of the other traditional tools and elements of a project, and if necessary, it can be the entire project plan. Similarly, just as the table of contents outlines the chapters and subchapters of a book, reading the table of contents first gives an idea of the flow of the book from beginning to end (like a project map), thus guiding the reader to the necessary and expected conclusion.

LEADERSHIP AND POWER: A PLANNED RELATIONSHIP

An organizational relationship has always existed between leadership and power; however, over time there has been a distinct change in this relationship as communication and communication systems have changed. Organizations of the distant past were limited by the physical proximity of the leader to exert power over others. Their ability to exert their command and control was critical within these organizations (Jaafari, 2006). Leaders had to be physically present, or loyal vassals of the leader had to be physically present to exert control. This form of leadership exists even today; there has been a transition from leaders who had to be present to leadership that could exist virtually. Leadership of virtual organizations has existed as long as humans have had the capability to communicate over great distances. The telegraph was a great innovation when introduced, but it lacked the robustness and flexibility of the wireless. When Guglielmo Marconi (1874–1937) developed the wireless, it allowed for many individuals to begin communicating over great distances as well as allowing people to function together as virtual teams.

Just as radio, movies, and television have revolutionized communications, future technologies will create a new world order of leadership and power within organizations. Organizations are the repository for leadership and power, and organizations define the relationship between the managers and the managed. Organizations with these elements have grown due to expanded communications potential. As communications have grown to accommodate larger and larger organizations, organizations have grown to meet this demand. Additionally, computing power has grown exponentially to meet these expanded needs. Data transmission has also improved in scope and size, accommodating larger transmissions of data. Organizations have continued to grow in order to utilize this increased bandwidth. Despite being able to transmit and receive larger chunks of information, the one area that is lagging behind is in the area of cross-cultural communication. One can expect to find in the future the ability of organizations to harness the ability of multilanguage transmission.

> Practical Tip: Consider how language might be affecting your team communication. Keep in mind that the language that you use might not be the primary language of the rest of the team. Sometimes it is better to pick up the phone and call to explain something than to send an impersonal e-mail. Consider your actions and make sure to use the best communication mechanism available.

Leadership and power has always been limited to the distance of communication, but the world has shrunk to create one large global village (Tichy, 2004). What experts fail to note is that we all live in this wonderful global village where were do not share a common language. While the means of communicating is available, the language skills have not quite caught up. While translations are available for many notable literary and musical works, the ability to offer real-time, global translations to every individual on the planet has yet to arrive. Once data transmissions have exceeded the realm of the needs of the reasonable user, then the additional computing power could be harvested to conquer challenges such as resolving language barriers that inhibit real global organizations. It is one thing to be able to communicate though a mutually agreed upon language; however, this limits individuals to languages that may not be their native tongue.

So the question could be posed, why would such immediate and far-reaching translations be important to leadership, power, managers, and the managed? All four of these elements are critical to organizations. Organizations are becoming increasingly challenged due to communication issues. Organizational culture is difficult to translate when one lacks the perspective of the society in which the organization exists. U.S. managers have learned the hard way that just transplanting good old American ideals and U.S. corporate culture to other lands is about as effective as harvesting your crops with a flamethrower. Global organizations thrive when they harness the potential of the local culture, rather than when they attempt to transplant the parent culture in another land.

When examining the elements of leadership, power, managers, and the managed, one finds that all of these considerations require effective communication. Without communication, all four of these elements are simply ineffective at best, and at worst they cease to exist. Leaders who cannot communicate with their followers will not be able to perform effectively. If communication is rudimentary, there are greater limitations upon the tasks the team will be able to complete. The power relationship is often more about the communication relationship. A leader must understand that the more rudimentary the form of communication available to the group, the greater the limitation upon the tasks that the team will be able to perform. Just as in war, where the radio gives the individual soldier the orders necessary to complete their task, communication gives the leader the ability to have the individual worker complete a task. If communication is more robust, then more complex tasks can be communicated. Good communications occurs when the message is transmitted in such a manner that the barriers to communication are marginalized, and the redundancy of the communication is increased.

Communication studies have shown that the actual words of the communication account for only a small portion of the message that is transmitted. When one examines the complete communication package, one finds that tone, body language, and demeanor account for the majority of the communication package. If one is not versed in the different cultural norms, when one communicates with someone from another culture, then tone, body language, and demeanor are almost always going to be incorrect for the cultural setting of the communication. Furthermore, the actual words of the communication, which account for the minority of the message, probably are not the native tongue of one or both of the people communicating. Other studies have also shown that when people are communicating in stressful or dangerous situations, they almost always revert to their native tongue for immediate communications. In times of stress, people have a predisposition to communicate in their native tongue. Thus, is it any wonder that global organizations in distress have difficulty communicating?

This limitation of communication has restricted leadership, power, and the relationship of managers and the managed. Once this restriction has been lifted and native, real-time translations are available to individuals that could interact virtually, this process would then alter organizational growth. No longer would organizations be limited to their span of control communications; it would then allow for organizations to expand at a growth rate before unseen in human history. Once organizations begin to expand into the multicultural, multilanguage dimension, the first benefactors of this growth will be those who can harness virtual teams.

The first place this translation technology should be available should be through virtual teams. Virtual teams are already limited in their communication, via the medium of exchange. However, early translation technology will allow virtual organizations to communicate with others while maintaining their virtual communication in their own native tongue. For example, as bandwidth and data transmission technology is increasing in power, simple voice communication through the Internet will have the capability to harness unused processing power to power translations of transmissions in this medium. Text transmissions could be translated instantly from short, direct text communications, while bundled e-mail, including text, could be translated en route to the native language of the receiver of the e-mail. This cipher-link meta-program would then put all people on equal footing in multicultural, multinational mega corporations. The early adapters of this technology would allow individuals to expand their corporate dimensions into areas and complexity previously not considered. A leader's power would increase, as a leader would then be able to manage greater numbers of diverse individuals. Key organizations would be able to expand past traditional national boundaries without the traditional exporting of personnel, culture, and processes. Other organizational processes could be controlled without adaptation and limited learning. Organizational knowledge would no longer be limited by language or culture. Organizations would be free to grow larger and stronger from the diversity of the workforce, as well as the enrichment that would ensue from the multicultural interaction. Additionally, this interaction could then also create more effective virtual teams that would improve organizational life, as well as expand cultural awareness. Multinational teams could then communicate with fewer communication inhibitors. Eventually, over time, virtual teams would begin to encompass other technologies that would further bring them toward traditional teams that exist in close proximity to each other.

In conclusion, technology is in transition, and teams and organizations are evolving. If leaders and power work best when there is physical proximity, then the relationship of the managers and the employees must utilize technology that will simulate physical proximity. When technology can replicate physical proximity, then teams, power, and the relationship between leaders and followers will have gone full circle.

CREATE A PLAN TO ADDRESS COMPLEXITY IN A VIRTUAL PROJECT

Make a Plan

In order to mitigate conflict, a PM must take steps to clearly define expectations, expected tasks, deadlines, and ramifications. The better individuals understand their role, the less likely that they will clash with others. Additionally, when everyone is aware of what is required of each individual on the team, there is less of a concern that some people are performing below expectations. When tasks are agreed upon up front, then there are fewer points of conflict due to ambiguity. A strategy to address this is for the PM to use a project plan to manage the virtual aspect of the team, as well as creating a project plan for the team goal. The PM in effect has to manage two projects while leading a virtual team. The virtual team PM should also plan to build a timeline and project plan for the project as well as one for the virtual environment. By utilizing two separate documents, it helps outline the process and expectations of the virtual team, but it also creates a performance document where team members can anticipate potential problems in the virtual environment. This document will also help keep the goals of the team in focus while explaining the challenges that might arise. Team members can then become committed to the project as well as being accountable for their actions.

> Practical Tip: Building a plan that addresses the needs of the virtual aspect of the team is as easy as putting together a shopping list. One needs to put together a list of all the things that need to be addressed and then make sure that those elements are discussed by the team.

Communicate the Virtual Project Plan as a Tool to Leverage Complexity as a Solution-Driven Process

Once the two plans are created, then both of these plans can be communicated to the organization to help fundamentally explain how the project will affect the organization. By addressing and anticipating conflict, it will prepare the organization for the rising tide. The project will always have periods of strife where outside forces will try to change the momentum of the project. Too often, uniformed outsiders try to redirect the project to meet different organizational agendas. This causes friction within a team. By offering a virtual plan anticipating potential problems, it addresses the questions posed by the uniformed outsiders. If the PM can continually keep these forces educated to the process, it can mitigate unwanted intervention by these outside forces.

In fact, the PM should feel that they are overcommunicating the plan. The reality is that repetition is one of the best forms of remembering, so the more the plan is communicated, the more likely that something will be done about it. Some might say that information overload might ensue, but there are far more commercials from five years ago that we can remember than the names of people with whom we worked five years ago.

If the PM has advised the organization that things might get rocky at certain points in the project, it will help reduce the level of organizational stress placed by the project. The project does not become a problem but becomes part of the culture. This cultural integration creates a more positive atmosphere for the project. It also keeps the project team on track. A leader has to remember that we live in an irregular world so not everything can be expected to run according to plan (Hamel, 2007). Often times one must abandon the plan in order to save the project.

CHAPTER SUMMARY

The relationship between leadership and power is important to understand because a leader can spend time increasing his or her power or giving power away. This should not be confused with delegation, which is extending a leader's power. Leaders must understand the importance of this relationship in order to be successful.

When a leader utilizes power in a planned manner, the leader is increasing his or her power. Power is important because it is one of the tools of project managers. Power can come from title or from expertise. Either type can assist a project manager move the project toward success.

The project manager must make a plan in order to anticipate the future of the organization. The plan must detail team formation, the potential for team conflict, and the support of the organization. The more the project manager can offer an expectation of the future, the better the team will be prepared to deal with the issues of the future.

The project manager needs to formulate a good communication plan from the start of the project. Effective communication about the project to all stakeholders will improve the likelihood of project success. If the project manager can keep the stakeholders aware of progress through good communication, there is a great chance of ultimate project success. If communication is good, there will be others who can step in to offer assistance if a project starts to fall into trouble. Without solid communication, the project can fall further behind because no one is aware. Intervention can only take place if people know that the team is in trouble.

REFERENCES

Brown, S., & Eisenhardt, K. (1997). The art of continuous change: Linking complexity theory. *Administrative Science Quarterly, 42*(1), 1.

Godin, S. (2008). *Tribes*. NY, NY: Penguin Group.

Hamel, G. (2007). *The Future of Management*. Boston, MA: Harvard Business School Press.

Hass, K. (2009). *Managing complex projects: A new model*. Vienna, VA: Management Concepts.

Jaafari, A. (2003). Project management in the age of complexity and change. *Project Management Journal, 34*(4), 47–57.

Krajewski, L. & Ritzman, L. (1996). *Operations management, strategy and analysis* (4th ed.). Reading, MA: Addison-Wesley Publishing Company.

O'Toole, J. (1996). *Leading change: The argument for value based leadership*. New York: Jossey-Bass.

Project Management Institute (Ed.). (2008). *A Guide to the Project Management Body of Knowledge—Fourth Edition*. Newtown Square, PA: PMI.

Rigby, D. (2009). *Winning in turbulence*. Boston, MA: Harvard Business Press.

Tichy, N. (2004). *The cycle of leadership: How great leaders teach their companies to win*. NY, NY: HarperBusiness.

Titcomb, T. (1998). *Chaos and complexity theory*. Landisville, PA: ASTD.

Chapter 6

Virtual Leadership through Complexity

Figure 6.1 Note all the elements of the picture: the water, the buildings, the boats, the sky, the trees, the bridge. Leaders must learn how to observe the details of a situation and to find the best path to move forward.

TEACHING LEADERSHIP SKILLS TO TEAMS FOR HANDLING CONFLICT IN A VIRTUAL TEAM

In order to lead through conflict in virtual team, it is a good idea to review some of the different points between a traditional team and a virtual team. Virtual teams must overcome many obstacles that face-to-face teams do not have to contend with. These issues include the challenges involved with the virtual environment, limitations of technology, and the lack of social cues associated with body language. Many virtual team members are new to one another, or they are new to working with each other in a virtual environment. A team leader must keep in mind that not all virtual team members will have similar levels of virtual experience. The virtual team depends on technology to communicate, rarely or never meets face to face, and the team members themselves make project decisions.

The lack of body language and visual cues in the virtual environment increases the potential for conflict. The team leader must monitor technology to ensure all the virtual tools are available and working properly because issues with technology will become an excuse or inhibitor to the success of the project. The virtual environment requires communication be clearer and more concise. Technology cannot replace poor communication; in fact, effective communication is one of the most critical elements of a virtual team. The leaders of the organization must support the virtual team by providing the needed resources for success. Without these resources, it will be difficult to achieve success. To overcome these challenges of the virtual environment, many studies and documented practical experiences emphasize the need for trust in the virtual environment. Trust is an essential aspect of virtual team success. It will not only help when things are going well, but it is essential when there is conflict.

Trust should become the foundation of a virtual team and studies indicate that without trust a virtual team was more likely to fail (Lipnack & Stamps, 2000). Trust is an integral part of successful virtual team leadership. Virtual leaders who have earned the team's trust will be able to navigate through conflict. The virtual team's leadership style benefits from establishing an effective way to promote trust and collaboration in a faceless environment. A leader must learn how to build trust given the challenging circumstances of the virtual environment.

A virtual leader should strive to leverage opportunities, support hierarchy, and have an initial face-to-face kick off meeting. These techniques can assist a leader in building trust. Furthermore, after building this trust, the virtual leader must learn to maintain trust by continuing to lead the group.

Creative manners and opportunities help the leader establish trust. A leader must learn to take counsel from others and to leverage opportunities when they arise. Often opportunities are short lived, so it is essential that a leader take advantage of these circumstances. Studies have shown that trust is essential for the success of a virtual team. Unlike a traditional team where fear might motivate some, trust is essential for a virtual team.

Hierarchy must be supported in order to maintain the team. Everyone must understand who their leader is and who to go to when they need assistance or advice. Without this element, the team will break down and team members might flounder, or problems that arise might not be forwarded to the correct people. Leadership must be present and available in order to maintain this hierarchy. Without the presence of leadership, individuals might not feel the need to report such issues. A virtual leader must also understand that from time to time a networked team will share leadership. This might break with hierarchy; the team must understand that this shared leadership is to manage tasks and not the process. Ultimately, the PM is responsible for the project, but there are times when the PM might delegate the authority and team leadership to someone who is a subject matter expert. This leveraging of internal team resources, as discussed regarding trust, is another way a leader can create an atmosphere of trust. A virtual leader must have a person with needed skills into a role of temporary leadership, when the skill is needed, in order to better assist the team. This creates a network team that is linked vertically through hierarchy, and horizontally by skill.

Another trust-building strategy is to have an initial face-to-face team meeting. Many virtual teams never meet face to face; studies suggest an initial physical meeting helps to increase trust, although it is not required. Even if an initial face-to-face meeting is not possible, the team still must meet together to start the process. However, understanding that an initial face-to-face meeting might assist in future success, a team leader should carefully review the possibility of an initial team meeting.

> Practical Tip: Trust is built over time and does not occur from a onetime meeting about trust. Make sure to talk about trust and to emphasize it to others regularly. A person serious about trust will talk about trust, just as a person serious about education will talk about educational achievements. Be a leader who is serious about trust by talking frequently about trust.

COSTS OF NEGATIVE CONFLICT

Negative conflict results in both wasted time and lost productivity for a project because conflict costs the organization money and is a contributing cause of project delays. It is estimated that the cost of conflict, when including ineffective managing of interpersonal situations, conflict avoidance, and lost project days, accounted for $20,000 per employee, per year. In addition, 20–25 percent of a manager's time may be spent dealing with team disagreements (Duarte & Snyder, 2006). The virtual project manager must identify potential periods of conflict as well as understand strategies to cope with destructive conflict.

Conflict can happen in any team at any time, so the prudent team leader must anticipate that there will be conflict within his or her teams (Duarte & Snyder, 2006). Sometimes conflict can be constructive, such as the conflict that comes when seasoned groups establish their roles within the team. Sometimes conflict can be destructive, when team members do not get along and actively undermine each other's efforts. Virtual organizations are particularly prone to conflict because team members lack the robust face-to-face contact that strengthens team bonds. Since virtual teams do not always have the luxury of having face-to-face meetings, it becomes the responsibility of the team leader to facilitate team building and conflict resolution within the group. Team leaders must be prepared to react quickly and decisively in order to resolve conflict and eliminate the productivity draining reality of conflict avoidance.

> Practical Tip: If you ever experience a negative cost of conflict, make sure to document the costs to communicate at the end of the project. This will help the organization to understand the cost of the conflict, as well as to keep it in mind for the next project. Make sure to highlight how the conflict was addressed and how the project was put back on track.

TIME AND PERFORMANCE MANAGEMENT FOR VIRTUAL TEAMS

Virtual leaders must understand the importance of time management, individual and team performance management, and communication style in order to successfully manage conflict. Since virtual teams are more geographically dispersed, as well as having to contend with time zone challenges, virtual leaders must adapt new techniques for success. By reviewing their existing resources in a new light, they can transform potential limitations to potential advantages.

Virtual leaders must learn to better leverage their time in order to keep the project on track. Traditional time management techniques can assist in this area. There is one important distinction to the virtual environment: virtual teams require fewer managers. Upon first review of this statement it would appear that fewer managers could cause a project to fail due to inadequate resources. However, a virtual team leader understands that in many cases less management is better management. In fact, leadership is more important but a wise team leader will understand that leadership does not rest in his or her own role alone. Virtual teams can better utilize shared leadership, where every individual is required to perform some role of leadership. Leadership will shift according to the requirements or objectives of the organization. A good virtual team leader should accept the ad hoc leadership that develops throughout the project (Duarte & Snyder, 2006; Lipnack & Stamps, 2000). By leveraging these situations, rather than be threatened by them, a virtual team leader can multiply his or her effectiveness. Allowing each virtual team member to carry the burden of leadership occasionally, even if only for a short while, will certainly allow the team leader to be more effective.

The performance management of virtual teams must become a consistently applied element of the project. The importance of timely team and individual feedback cannot be underestimated. A common belief is that virtual team members can be left alone. However, the reality is that team members require clearer performance measures than colocated teams. When virtual team members are left alone, they feel disenfranchised and apart from the team. A good virtual team leader must create a consistent measure of individual and team performance in order to better manage the results of the individual and the team. This continuous feedback is important because it keeps the project on track as well as keeping individuals focused on the goal(s) of the group. This method is important for avoiding conflict, but can also be used to correct a situation once a conflict arises. Feedback and coaching sessions can be used to work through any problems or conflicts that arise.

In summary, conflict is inevitable. A virtual PM must be prepared to address these problems when they arise. They should be addressed as soon as they come to the attention of the leaders. Creating a culture that addresses conflict instead of ignoring it is the single most important element of handling conflict. If the leader's modus operandi is to ignore conflict, then others will ignore conflict until it explodes. Leaders must learn to utilize strategies to reduce conflict; leaders must address conflict; and leaders must anticipate conflict. All of these are essential to create a culture that handles conflict. Without these strategies, a virtual PM may allow a situation to fester and eventually undermine the entire project. Feuding team members are worse to a project than lack of resources because a dispute will cause team members to recklessly allocate resources to support their own agenda, instead of supporting the project.

> Practical Tip: Plans are the walls of life, the silos that keep us from understanding others. There are always choices to be made, so one should plan to try to understand another person's perspective in order to become a better leader. Sometimes thinking about how another person feels will lead one to understand that person better.

CASE STUDIES OF COMPLEXITY THEORY APPLIED TO SUCCESSFUL PROJECTS

Case Study 1: Reducing Conflict within an Organization

Groundmoving Industries (GI) is a manufacturer of diversified heavy machinery used in the road-building industry. GI is one of the top manufacturing companies of asphalt plants in the U.S. and Brazil. Asphalt is the primary product used in new road manufacturing as well as in maintaining of existing roads. GI current global economic goal is to expand their market into Canada. GI has been highly successful with the marketing of asphalt manufacturing plants and has a good reputation and a high level of product reliability.

GI has manufacturing facilities in both the United States (Long Beach, California) and in Brazil (São Paulo). The CEO of GI, who is based in California, has stated that the global objective of the company is to expand into Canada, there has been a virtual conflict going on between the office in Long Beach and the office in São Paulo. The expansion into the Canadian market is clearly targeted at benefiting the Long Beach operations. All the senior management has agreed upon the strategy it was clear that the senior management in Brazil went along with the strategy based upon the clear evidence of the soundness of the strategy, rather than the belief that it would benefit the São Paulo office.

When examining the decision of companies to purchase large heavy equipment, freight and shipping is clearly a costly issue to be taken into account. GI supplies machinery that is large and heavy, requiring complex freight orchestration. GI manufactures equipment that is extremely bulky and is best and cheapest to be transported along roads at certain times.

The equipment manufactured by GI is so heavy that transportation via conventional container ships is more costly. The closer the client is to the manufacturing facility, the less the cost of transportation. GI is based in California and has a strong relationship with a national freight company proficient in handling the local permitting required for transportation by truck. The U.S. market is based upon expansion and reworking of existing highways. In most other countries GI equipment is used to build more new roads or to expand existing roadways. This means that there is less of an infrastructure to transport this large and heavy equipment, so transport in other nations requires combination of truck and ocean freight.

There are four clear reasons for this strategy, which have been discussed by senior management. However, these have not be disclosed to the entire organization because the company feels that this kind of market information is treated very carefully since this kind of information could cause their competition to shift toward this market, creating unwanted competition in this new arena. The first of the four reasons is the current growth of the Canadian market. The second reason is the high level of market potential in Canada. The third reason is the current exchange rate between the U.S. dollar and the Canadian dollar. The fourth reason is the environmentally friendly aspects of the GI product.

Canada's GDP is growing. Currently, Canadian GDP growth is just under 4 percent, but it is expected to rise to 5–8 percent or greater in just a few years. In contrast, U.S. GDP growth is expected to fall to 1.8 percent for the next several years. This weakening of the U.S. economy and the strengthening of the Canadian economy will benefit GI. If the relative economies change in the directions predicted by the experts, it will place GI in an excellent export position. This will also occur at a strategic time; with U.S. GDP growth slowing to 1.8 percent, this would also mean decreased domestic consumption and purchases. GI hopes that this future expansion will help supplement the decreased consumption domestically.

Information has shown that almost 40 percent of Canada's highways are not paved or will require repaving in the next few years. Furthermore, there is also a huge potential market in repairing existing paved highways. GI has clear market potential in Canada when examining the need for heavy machinery in the manufacturing of asphalt (the primary element of paved roads).

Foreign exchange is another critical factor in the export expansion of GI into Canada. The Canadian dollar is in relative parity with the U.S.

dollar. Furthermore, GI has put emphasis on its brand image of quality and reliability necessary for a product of this nature. GI can market the extremely low downtimes of such equipment, and offer to display its reliability by testimonials from satisfied customers. The building of a brand will begin to generate interest in the product, while creating the need in the mind of the consumer. If GI structures its message correctly, GI will begin to get requests that culminate in orders once the financial requirements shift to support the wants of the market.

GI produces more environmentally friendly products that use recycled materials in manufacturing than the competition does. Since Canada has similar environmental laws to those in the U.S., this would allow GI to highlight its commitment to U.S. standards. GI would stand above other international competitors whose environmental requirements may not be as stringent as those in the U.S. and Canada.

There are clear economic reasons for the company to want to export their products to Canada; this is clearly causing friction between the two organizational offices. Given the information presented in this case study, explain what could be done to better reduce the conflict between the two offices.

- What changes in communication would you recommend in order to reduce the conflict between the two offices?
- What are some of the costs of this kind of negative conflict within the organization?
- What could be done to improve the trust between the two offices?
- What kind of performance management could be applied to this situation?

Case Study 2: New Virgin Group Communication

The Virgin Group™ has been highly successful in inventing themselves in multiple, diverse markets. By 1992 Richard Branson had created a $1 billion dollar record empire with his company, Virgin Records™. Virgin Records™ was the largest independent label and included in its portfolio such stars as Janet Jackson, Phil Collins, and the Rolling Stones. This firm has become the model for the successful strategy of global marketing and strategic planning. The Virgin Group™ is a diversified conglomerate spanning 150 different private companies worldwide. The reason for this towering success has been due to Branson's ability to apply global marketing techniques, and to his critical timing in the area of strategic planning.

The Virgin Group™ (VG) is a privately held conglomerate of companies with revenues around $2 billion dollars. The Virgin Group™ is the largest U.K.-based private company, with assets primarily in the travel, entertainment, and retailing sectors. In the travel sector, Virgin Atlantic™ is the shining airline upstart. Virgin Atlantic™ has taken on British Airways and has been successful primarily due to its ability to create what is known as "entertainment at 20,000 feet." In the entertainment area, V2, Virgin's reentry into the music industry after the sale of Virgin Records™ to Thorn-EMI, is positioning itself as an independent record label. In the retailing sector, Virgin Cola™ has become the number two soft drink in the U.K., usurping the position previously held by Pepsi. VG continues to expand; however, we will now review how a strategy of global marketing and planning has supported this company to its current level.

Richard Branson is VG. The global marketing position starts and ends with the larger-than-life personality of Richard Branson. The charming and witty media star makes everything he does a media event. From the attempted balloon trips around the world, to the BA Bonus distributed to Virgin Atlantic™ employees, everything he touches must be an event to be covered. VG is the company; Richard Branson is the brand. By marketing himself and his company at the same time, and by creating a media circus wherever he goes, Richard Branson has shown the power of a global brand. VG has a presence around the world and continues to expand.

This corporate synergy allows the company to build upon its own resources. For example, Virgin Bride (the chain of bridal stores that are part of VG) offers full planning of a wedding and honeymoon. What honeymoon would be complete without a trip on Virgin Atlantic™, and possibly a stay on Neckers Island (VG's private island)? This VG synergy allows for each company to leverage the strengths of the network. Strategic planning of firms that offer greater strength is part of the strategic vision. In addition, VG is also the master of timing. They know when to get into an industry and when to pull out.

Richard Branson and by extension VG has contacted you to develop a new business unit as part of the Virgin Group™. The concept is that they will build a new theme park in Dubai as part of The World chain of islands. Richard Branson has already purchased the island of Great Briton as part of The World island network off the coast of Dubai, and his development concept is to build a Virgin-themed entertainment complex there. Now that you have been assigned this project, you feel that your first step is to create a communication project plan to support the progress of this new project. You believe there is significant synergy within VG to make the endeavor even more successful by including the other business divisions.

- How will you arrange communication between the new Virgin theme park and the other divisions?
- What would be the plan of communication between you and Richard Branson?
- What other stakeholders might you also remain in communication with?
- Since you will be located in Dubai to oversee the development and most of VG will be in the UK, how will you address the virtual nature of your project?
- What might some of the tasks of the shadow project plan look like?
- What kind of communication will you have with the media and how will you channel that information?

Case Study 3: Leadership and Organizational Culture Case Study

Mergers and acquisitions have been a significant factor in business for the last seventy-five years. In particular, due to the recent changes in the economic climate, companies have been continually morphing via corporate mergers and acquisitions. Furthermore, in the last thirty years, there has been considerable evidence to indicate that the majority of mergers and acquisitions do not create value for the stakeholders of the acquiring company (Eccles, Lanes, & Wilson, 2001, p. 46). Despite this research, mergers and acquisitions remain a common aspect of corporate life in the United States. Mergers and acquisitions continue to infiltrate the business world, and the current trend indicates more will occur in the future. As organizations progress through their life cycle, and as these industries reach maturity, these companies become ripe for consolidation. Plenty of information has been generated about the relative financial successes and failures of mergers and acquisitions; there has been less of a focus on the human factors related to these changes. Too often shareholders focus solely upon the return on investment and fail to quantify the effects of leadership and organizational cultural facts involved with these deals.

This case study will review the importance of the effects of leadership and culture upon a merger to explain why these elements must be emphasized in order to lead to financial success.

The Daimler-Chrysler (DCX) merger will serve as an instructive example of what can happen when leadership and cultural issues are not given a high enough priority by the merging organizations. By examining the genesis of the merger, one can see the various challenges that occurred throughout the process. The DCX merger is of particular interest as it began as a merger of equals, with leadership focus upon cultural factors; however, within a short period of time it was apparent that it was a takeover and then the clash of cultures occurred. After years of struggling, Daimler finally parted ways with Chrysler, with the end result being that Cerberus Capital supplied the funds to create a reborn Chrysler LLC.

At the time Daimler-Benz and Chrysler were considering a merger of their organizations, the world's vehicle manufacturing industry was undergoing a rapid consolidation. The factors driving the consolidation were predominantly globalization, excess capacity, and the need for economies of scale and the desire for improved profits. Furthermore, of the top 20 automakers in 1965, 14 have since merged or been taken over (1998 Automotive Deal Survey, p. 7).

Leadership at the Time of the Merger

Jurgen Schrempp, Chairman of Daimler-Benz Jurgen Schrempp is an autocratic type of leader. This is not always a bad style of leadership, but it can lead to issues if the organization's culture is not receptive to this type of leadership. Authoritarian leaders must coerce and persuade others, and they must leverage the power of their position, or they must utilize their power to punish or reward individuals to continually perform the required behaviors. Authoritarian leaders must dominate their followers so they will adhere to procedures. Compliance through coercion is the authoritarian leader's modus operandi.

As the chairman of Daimler-Benz prior to the merger, Jurgen Schrempp has built a reputation as an autocratic leader who orchestrated a dramatic turnaround at Daimler-Benz (*BusinessWeek,* Jurgen Schrempp, 1/11/99). He turned their 1995 operating loss of $3.45 billion around to become profitable by 1996 (*BusinessWeek,* 11/16/1998). His self-described management philosophy is based on what he calls the "Schrempp Curve," whereby decisions are planned and implemented in small increments. Based on Schrempp's philosophy, leadership's confidence in decisions must remain unwavering, in order to instill the trust of the entire organization, in a step-by-step and methodical manner (*BusinessWeek,* 11/5/1998). His emphasis is on decision-making speed and implementation. His priorities for DCX include profitability, attractiveness to employees, and good corporate citizenship, in that order (*BusinessWeek,* 11/5/1998).

Robert Eaton, Chairman of Chrysler Corporation Robert Eaton, chairman of Chrysler Corporation at the time of the merger, viewed the priorities for the merged company a bit differently. His more collaborative style of leadership was evident in his approach to the DCX merger implementation plans. According to Eaton, the integration of the two organizations and subsequent development of a new culture were his top priorities. He recognized that synergies could result from the combination of organizational resources and predicted that these synergies would help the organization achieve its predicted $1.4 billion in cost cuts for the first year of their business together (*Automotive News,* 11/9/1998).

Given the differences in the leadership of these two organizations, it would appear that the leaders should have had more of a plan to navigate the uncharted waters of their company's future, while internally wrestling with their own bureaucracy. Leaders must move an organization forward, and they must chart a course that others will want to follow. In the case of a merger, such as DCX, one would have predicted that Eaton and Schrempp would have combined forces to operate as an effective team at the top, with each giving leadership on different sides of the Atlantic. Instead we quickly found that Daimler and Schrempp had a different view, and it became obvious that Schrempp was a hard-core turnaround leader with no time to worry about teams (Katzenbach, 1998, p. 158).

The Merged Cultures

The new organization started in a state of chaos and uncertainty. In fact, in the beginning of a merger or acquisition, there are two distinct cultures competing for supremacy. To state that initially there existed two distinct cultures would be to better describe the situation at DCX. Over time, the acquisition of Chrysler by Daimler has certainly changed the culture of both of their previously separate organizations. Ultimately, the culture of Daimler has certainly eclipsed the culture at Chrysler. However, this was not with problems and fall out. One of the early departures was of Robert Eaton, who left the company within a year of the merger. To better understand the culture effects, one needs to review the cultural obstacles that were before the leaders of DCX.

Cultural Obstacles

When one examines DCX, two factors are found that profoundly affect the culture of this new mega-organization. The first is the sheer size of the new organization. Financially, this new company is only one organization that does not instantly transform this company into one culture. In fact, the sheer size of this new organization alters the culture. There are always suborganizations with different cultural attributes that will emerge any time there is a large culture. In the case of DCX, Daimler is clearly the dominant culture; however, Chrysler remains a subculture of this greater organization. This would severely hamper the organization as well as lead to Eaton leaving the merged entity much faster than anyone would have predicted.

Second, when examining the affects of the acquisition, one finds the acquisition process changes the culture. DCX has been transformed through the acquisition process, making the assimilation of Chrysler a complicated clash of the two cultures. Despite the fact Daimler-Chrysler is now one unified economic entity, this does not immediately imply that it is one unified culture. Cultural factors have clearly affected this acquisition, and the clash of the different cultures has caused conflict and unrest within both organizations. This unrest and conflict has certainly affected profits and production.

Clash of Cultures

The subsequent clash of cultures clearly shows these two automakers do not agree on how they should achieve their ultimate goal. There have been other cases of U.S. companies being purchased by German organizations—the purchase of Westinghouse by Siemens being a recent example—and in all of these cases there have been significant cultural issues. In some cases the long-term return has justified any short-term loss; the DCX situation has yet to produce the anticipated results. It is clear that early on in the process, the leadership and cultural factors were considered unimportant; one can only monitor progress of the financial indicators to determine if Daimler gains significant value from this transaction, or if they will add their name to the growing list of companies that participated in failed mergers and acquisitions.

At the onset, the combination of Daimler-Benz and Chrysler was referred to as a "merger of equals." The leaders of both companies were attempting to fill the gaps that existed within each other's global positioning and capacity utilization, through the addition of each other's complimentary product assortments, manufacturing resources, and distribution systems. A great deal of strategic analysis was performed prior to the merger to conclude that Daimler had little overlap with Chrysler in the product area (*Business Week,* 11/16/98). Based on an assessment of both company's operational goals and strategies, there was no question the merger would benefit them both, from a sales and financial perspective.

Language and Cultural Issues

There were major differences between the corporate cultures at Daimler and Chrysler. The most obvious difference between the two entities is the language barrier, with Daimler-Benz operating as a German-based and -speaking company and Chrysler existing as an American-based, English-speaking company. Inherent in the foreign language obstacle is also the difference in international cultural etiquette. According to Steve Harris, Chrysler's former communications chief, Germans pride themselves on analytical research that produces a plan, while the American way is to try for the impossible and keep coming up with new ideas to make it happen (Daimler-Benz and Chrysler to Merge to Daimler-Chrysler, 7/8/99).

Human Resource Challenges

In addition to their language differences, they face a few human resource differences as well. While Chrysler employed 120,000 people, all in the automotive manufacturing area, Daimler-Benz employed 320,000 employees, many in their nonautomotive industries (Daimler-Benz and Chrysler to Merge to Daimler-Chrysler, 7/8/99). It was unclear at the time of the merger how many jobs would need to be cut in order to operate the new company most efficiently, yet both Schrempp and co-CEO Robert Eaton stated the merger of Daimler and Chrysler would not result in job elimination at any level (*Detroit News,* 11/15/1998).

Leadership Strategy

Immediately after the merger, executives of both companies bragged about the informal approach they took in planning and combining the

Daimler-Benz and Chrysler cultures, while priding themselves on the following achievements:

More than $1.4 billion in savings
Stock price of $94
Thirty-four new products already on the drawing board (Daimler-Benz and Chrysler to Merge to Daimler-Chrysler, 7/8/99)

Clearly, addressing the problems of an overabundant human resource pool was not perceived, nor communicated, as a top priority for the company's leadership. This area became a serious problem as the merger moved forward and the savings and value did not materialize. Leadership looked more at job reduction as a way to improve the financials of the company; however, in the end both Shremp and Eaton parted from the Daimler organization and the organizations took years to recover from the financial losses stemming from the merger.

The problems never left the Chrysler organization after this difficult time and after being bought by an investment group it appears that Chrysler will cease to exist.

Questions

- Offer your opinion upon the leadership at the time of the merger. Do you feel that this contributed to the ultimate failure of the merger?
- Which of the culture factors discussed did you find to be the most important in the failure of the merger?
- How might have the clash of cultures be avoided? What could have been done differently?
- If you were one of these leaders, Schrempp or Eaton, what would you have done differently? What steps would you have taken at the time of the merger and immediately following?
- Do you feel that most mergers or acquisitions offer the shareholders greater value, or do you feel this example is more typical of what happens?

Case Study Review

There are no clear answers for any of these cases, but consider these three key factors: trust, culture, chaos. In each of these cases trust, culture, and chaos were significant factors. First, these organizations lacked some degree of trust, which caused significant problems. Second, a clash of cultures is important to understand as a potential underlying factor in any business. Third, the more that chaos is allowed without restraint, the more that confusion can bring down a company.

Essentially, trust needs to be built up and to be a continuous improvement process in any organization in transition. At any given time, most organizations are in the middle of some kind of major transition, whether real or perceived. Wise project managers keep an eye open for this and do their best to continually build up trust.

Culture is an organizational concept that is continually being built and rebuilt. The more that an organization is learning, growing, existing, the more the culture evolves. Humans are social beings who need to communicate and the more bad news available, the more the culture will gravitate toward that information. A wise project manager tries to inject positive news into the organization or remind individuals of achievements rather than propagating more of the negative.

Unrestrained chaos is the bane of all organisms. Just like a virus that remains unchecked, that can ultimately destroy the host, unrestrained chaos is similar to a disease. Negative chaos can be combated by structure, order, and positivism. The more the organization can be perceived as on the right track, the less negative chaos will enter into the belief of the organization. Just as stock prices rise and fall by the whim of the masses, one needs to be wary of the challenges of unchecked negative chaos. Disorder leads to disharmony, disharmony leads to distrust, distrust leads to disorder. Hence, it becomes a vicious negative cycle that can only be corrected by significant effort or destruction of the organization.

Thus, consider these three elements in the previous case studies and consider how these elements contributed to the problems encountered. Even great organizations can have hard times, but it is a matter of how the problems are managed more than the problems themselves. No one faults an organization for the loss of business in a poor economy and taking the necessary steps to reduce costs, but everyone faults an organization for not laying off people with dignity.

CHAPTER SUMMARY

Some conflict is inevitable for any project. A virtual PM must be prepared to address conflict when it arises and it is best if the virtual PM has a plan to address the conflict that does arise. Creating a culture that addresses conflict instead of ignoring it is the single most important element of handling conflict. Conflict that is ignored can be toxic to a project and will certainly lead to lost time and lost productivity.

Leaders must be more careful when dealing with conflict in a virtual setting. There are fewer cues to observe the conflict so it may go undetected for some time. The virtual leader will need to make special efforts to make sure that virtual conflict is addressed. The clearer the communication, the better the conflict will be addressed. A virtual PM needs to find and address any conflict that arises in order to make the organization as effective as possible.

REFERENCES

Bob Eaton to retire following record year at DaimlerChrysler. (January 26, 2000). DaimlerChrysler – News, http://www.daimler-benz.com/news/top/2000/t00126a_e.htm.

Could merger mix have helped DaimlerChrysler? (5/9/ 2001). ManagementFirst.com, www.managementfirst.com/articles/daimler.htm.

DaimlerChrysler expects steady growth in key markets; Record sales and profit figures in the first business year. (March 31, 1999). DaimlerChrysler—News [On-line]. Available: http://www.daimler-benz.com/news/top/1999/t90331b_e.htm.

Duarte, D. & Snyder, N. (2006). *Mastering virtual teams: Strategies, Tools, and Techniques That Succeed*. San Francisco: Jossey-Bass

Eaton works to build "new culture" at D-C. (November 9, 1998). Automotive News [On-line], Galenet.galegroup.com: 112X4487.

Eccles, R. G., Lanes, K. L., & Wilson, T. C. (2001). Are you paying too much for that acquisition? In *Harvard business review on mergers and acquisitions* (pp. 45–71). Boston: Harvard Business School Press.

Jurgen E. Schrempp: Deal of the decade. (December 30, 1998). BusinessWeek [On-line], www.businessweek.com/1999/02/b3611006.htm.

Katzenbach, J. R. (1998). *Teams at the top: Unleashing the potential of both teams and individual leaders*. Boston: Harvard Business School Press.

Katzenbach, J. R. (2000). *Peak performance: Aligning the hearts and minds of your employees*. Boston: Harvard Business School Press.

Lipnack. J. & Stamps, J. (2000). *Virtual teams: People working across boundaries with technology* (2nd ed.). NY, NY: John Wiley & Sons.

Schrempp cocktail. (May 20, 1995). *Economist, 335*(7915), 62.

Schrempp Q & A: "We must be the greatest transport company in the world." BusinessWeek (International Edition) [On-line], www.businessweek.com/1998/46/b4604053.htm.

Schrempp's repair job. (March 3, 2001). Economist [On-line], EBSCOhost: AN: 4150900 ISSN: 0013-0613.

Shareholders vent anger DaimlerChrysler boss gets heat for 1998 merger. (April 12, 2001). Arizona Republic [On-line], Proquest: ISSN: 08928711.

Chapter 7

How Organizational Culture is the Key to Applying Complexity

Figure 7.1 The road seemingly to nowhere is an example of how it feels trying to learn more about complexity. The further one travels down the road, the more one learns and the more relevance the path will take until finally one connects with the journey.

INTRODUCTION TO ORGANIZATIONAL CULTURE

Culture is a system of shared assumptions that guide a group toward resolving its internal and external challenges. Further, it gives a framework to the group or organization in a manner that allows the system to interact, judge, and intimidate other external groups and internal subgroups. Culture is the web that binds individuals together in a way that they have a shared ethos (Schein, 2004). Understanding the framework of culture gives an individual a paradigm to apply when interacting with foreign cultures. Additionally, it is important to have a framework to describe and define culture so one can readily identify and classify new cultures (Schein, 2004). Cultural complexity is the understanding that shared assumptions offer a web of judgments, guidelines, and principles that define and guide a group. This web is not one of formal or hierarchical design but is one that is developed through leadership, experience, observation, and organizational lore.

The dilemma is whether a project manager should allow cultural complexity to take its course in a project or should resolve the situation. There is sufficient evidence to suggest a paradigm shift is needed for managing complex projects, which encompasses virtual projects, from the traditional project management processes (Cooke-Davies, Cicmil, Crawford, & Richardson, 2007; Singh & Singh, 2002; Leban, 2003). PMI does not suggest a formula for managing complexity. In the past few years, there has been an upward growth of projects as organizations continue to grow in size, companies continue to outsource, the world continues to see people cross borders, and the project manager must deal with the additional complexity of culture. The project manager is now faced with significant lines of communication because of the sheer number of individuals on the team. Furthermore, a project manager may be dealing with language barriers, ethical issues, technology dilemmas, and cultural tensions, among other issues that happen between people (Krajewski & Ritzman, 1996). Some of these situations can be dealt with through processes and procedures, but should they? Project managers must allow individuals to resolve their own problems in order to stimulate the development of autonomous work teams. Not all situations can be resolved with a simple yes or no answer, especially with the numerous cultures and technologies involved. The situation that needs to be settled may be resolved by several telephone calls rather than having a process and procedure in place. The project manager must review this after the fact. Certainly, where laws and regulations are a factor, this takes precedence and a process and procedure must be put in place.

LEADERSHIP AND ORGANIZATIONAL CULTURE

Shaping culture becomes the missing link in how to address complexity theory. As already shown, with a large and complex project or organization, the project manager cannot be present to address each and every challenge (Krajewski & Ritzman, 2001). Shaping a culture takes time and the project manager must infuse the culture with the vision of what the organization will look like in the future. Merging the vision with the culture and combining them to become one will allow the entire organization to move forward in unison.

It is more than just delegation or laissez-faire leadership; it is a blending of the old and the new. It incorporates more by including several new levels for project managers to utilize. For example, one can look at the structure of commando groups or terrorist cells—small autonomous groups that are able to enact great change without daily direct involvement by a leader (Marcinko & Weisman, 1997). This is the same kind of micro-organization that is imbued with a common culture that helps direct the group toward a mutual goal. Terrorist cells are clearly aligned with a negative purpose; one can examine the structure to better understand how they operate. Similarly, commandos operate upon analogous principles, but are still charged with waging war upon other groups as sanctioned by a government.

Complexity theory is akin to this type of autonomous organization. Applying this type of successful micro-culture to a project is a direct application of complexity theory. A leader must spend time as the ambassador of the culture and then allow the subordinates to develop their own path toward project success (Marcinko & Weisman, 1997; Schein, 2004). This goal can be misguided, as with terrorist organizations, or it can be applied for higher purposes. In any event, these micro-organizations are the visible application of complexity theory.

Virtual Project Management Monograph

Khazanchi and Zigurs (2005) were sponsored by the Project Management Institute™ (PMI) and the University of Nebraska at Omaha to study the patterns of effective management of virtual projects. This paper will review the findings of the study as they correlate to complexity theory and how they subsequently can be applied to culture from a complexity viewpoint.

The study divided the virtual projects into three groups. The projects were lean, hybrid, and extreme. Lean was defined as "low complexity, narrow scope, low risk" (Khazanchi & Zigurs, 2005, p. 30). Hybrid was "varying levels complexity, scope, and risk" (p. 30). An extreme virtual project was "high complexity, broad scope, high risk—e.g., mission critical" (p. 30). This paper will focus on the results for extreme virtual projects. There were a total of 17 participants from five different companies from several different Western countries.

Only those findings that had some correlation to complexity theory and cultural implications are reported here (See Table 7.1). The information listed in the table is verbatim from Khazanchi and Zigurs's monograph (2005). The information italicized is the beginning of complexity theory; and will be discussed in the findings portion of this paper. For the most part, most of the monograph's findings were the customary suggestions that have been traditionally used by practitioners.

SPENDING TIME DEVELOPING THE CULTURE OF COMPLEXITY

Lipnack & Stamps (2000) concludes that trust is a major component of a successful virtual team. This is supported by the studies done by Duarte and Snyder (2006, p. 21). To maintain trust, a leader of a virtual team should "set and maintain values, boundaries, and consistency" (p. 83). This becomes even more important in the virtual team since the team membership may be fluid and for some short-lived. To build and maintain trust with a short-term member is even more difficult, according to Duarte and Snyder (2006).

Transformational and contingency-based leaders both must deal with linguistic hurdles and other limitations. The transformational leader will leverage language to improve the team. A contingency-based leader is limited to clearly expressing ideas and concepts in a direct and concise manner. This is favored by some, and important to many, but it does surrender some of the richness of human thought and language. Two of the most powerful complexity-based cultural weapons are myths and metaphors. By combining these with indices, icons, symbols, slogans, and other relevant codes, the transformational leader can drive the team beyond the typical achievements. This use of complexity goes beyond a

Table 7.1 Extreme Virtual Project Patterns

Context	**Problem**	**Solution**
Team members do not feel that they are a unified whole. People feel they are working independently rather than together.	How do you create synergy in your team and a shared understanding of project goals?	Use face-to-face or video conferencing to introduce and socialize team members at the inception of a project. Communicate clearly and often on project goals and individuals' roles in the project. *Create a culture that encourages the sharing of issues, good and bad news, and all project-related information; a discussion of solutions; and the flexibility to accept differences.*
Your team members are unclear about their roles and responsibilities in the project. This is causing misunderstandings about project goals and resulting in a delayed project.	How do you provide team members with a clear understanding of their individual roles and responsibilities in the project?	Clearly define team members' roles and responsibilities and work processes at the outset. If new members are added, clearly communicate revised roles and responsibilities along with timelines and tasks to all the team members. Ensure that they all understand their assignment and provide them with the tools to deliver. Communicate roles, responsibilities, and work processes to all stakeholders and team members. *If feasible, consider rotating members through different roles.* Use technologies with a high process structure (such as virtual collaboration systems and knowledge management tools) to share information on the team's work processes and roles/responsibilities of team members. *Include team members in designing work processes and delineating roles and responsibilities.* This will increase team ownership.

(continued)

Table 7.2 *(Continued)*

Context	Problem	Solution
You are unable to fix something within the team, and progress is stalled until this issue is resolved. This is difficult enough in a colocated office environment, but the virtual nature of your project substantially increases the magnitude of the problem/issue.	How do you coordinate access to human expertise to assist with problems that are stalling the project?	Plan for, and *provide the project team easy access to, subject-matter experts, technical experts,* and experienced software/system architects who can provide advice/help needed to resolve problems/issues that crop up during the project.
Your team is having difficulties accepting each other's information. This is aggravated by the difficulty of building mutual trust in a virtual setting.	How do you build strong relationships among team members?	Set up immediate trust between team members while working on building mutual trust over time. *Encourage a team culture that accepts constructive criticism and questioning of each other's deliverables. A team culture that builds on respect, courtesy, open communication, and flexibility will go far in building stronger relationships.*
Your team is not able to manage conflicts and resolve them during discussions on a shared understanding of requirements, resolution of project-related issues, and challenges.	How do you establish consensus and resolve conflicts effectively when team members have differing personalities, cultural and language backgrounds, and personal goals?	Establish and communicate the project vision, goals, tasks, roles, and responsibilities clearly. When team members have shared goals and work toward them, projects are successful. At the outset, establish, in consultation with the team, a clear work process for evaluating recommendations on issues and handling conflicts. *Potentially, you could use a team steering group to handle conflicts that do not reach a consensus solution.*

Your team does not have adequate project and technical leadership.	What are the elements of effective project leadership?	Effective project leadership in virtual teamwork is absolutely essential for project success. It is achieved when there is a single individual tasked with leading the virtual team, and when this person is held accountable for both success and failure. *The leader develops (in consultation with the team)* and credibly communicates the project vision and strategy. *This helps create a culture of open communication and consensual decision-making.* The leader also motivates and inspires team members so that they are energized and are able to overcome any barriers of working in virtual settings, including time zone and geographic differences, organizational and political challenges, and resource constraints. Finally, an effective leader is also responsible for aligning the "right" people for the project and tasks therein.

contingency-based deterministic and mechanical model of projects and elevates it toward a more social and cultural model.

Myths and metaphors are expressions of the complexity of human systems. When used correctly, both myths and metaphors can be used to move an organization toward greater success. Transformational leaders can communicate workplace experiences in a manner that creates a social lesson and instills values in the process. These myths and metaphors can be used to leverage a team or any group of stakeholders toward a higher level of change.

An organizational myth can become a central theme of a project in order to help transform it beyond the confines of the actual project. There are common myths about people and actions that transform deeds into acts of greatness. This is unique to human systems; myths are more than the sum of their elements (Carin, 1997). In society today, most of us know the myth of Johnny Appleseed and the associated lessons that come with this myth (Osborne, 1991). It is not clear if Johnny was really a person or group of people, but the myth creates an ideal where a person can achieve greatness, immortality, and fame by actions that could be described as ordinary. What is interesting about this complexity-based social phenomenon is that the message in the Johnny Appleseed myth is the same message as the *butterfly* example of complexity. The flapping of a single butterfly (or the actions of a single person) can create great change or effects elsewhere or later in time.

Another myth to review is directly related to the complexity-based organization. The myth about John Henry is one that is full of metaphors for complexity. On the surface, the myth about John Henry appears to be nothing more than a folktale of a man achieving greatness through personal work-related deeds; however, when one examines the metaphors and symbolism in the myth, there is much more at work. The John Henry myth is about a healthy and strong black man, who has been used to heavy labor for his entire life, challenging a steam-powered machine to see which can hammer faster (Osborne, 1991). There is some evidence that there was a real man named John Henry who worked on the C & O railroad about the time the myth is reported to have taken place. Yet John Henry is more of a symbol of workers at the time.

During the time of the myth, which is approximately 1870 and takes place in Tallcott, West Virginia, there was increasing pressure upon human laborers in many different industries as these individuals were being replaced with mechanization (Osborne, 1991). Mechanization and industrialization had been occurring for some time in the United States; during this time there were particular pressures upon different individuals in society. The United States Civil War had just ended five years prior to this myth; the end of the war declared all slaves forever free, and society had been coming to grips with integrating these newly freed slaves into a wage-based economy (Osborne, 1991). There was plenty of rebuilding to do after years of war, but there was not always a lot of money available. Furthermore, social pressures were reducing certain jobs as the economy moved to repair itself. This myth has layers of metaphors and a richness of understanding that can only be possible in human society.

This myth is the great man versus machine matchup that has caused society to ebb and flow about whether a holistic or a mechanistic understanding is better. What is interesting about the message of the myth is man (John Henry) does beat machine, but cannot celebrate his victory because he dies shortly after the end of the contest. An interesting message of the story is that man is clearly superior to machine but the machine has no soul and cannot die (Osborne, 1991). John Henry, who was by all accounts one of the strongest men of the time, defeats his nemesis but dies in the process. It does show that an extraordinary man can still defeat the progress of technology, but he is still mortal. This reflects the change in 1870s society where human labor is being replaced by machines for many labor-intensive jobs. Not only does this happen in 1870, but machines replace humans again and again in history. Part of the reason that this myth has persisted for so long is that the message is reoccurring in society. It becomes its own complexity-fueled story that grows in the telling.

So, given the longevity and messages of these two myths, one must bring this back to transformational leadership and complexity. We must ask ourselves, is there room for another Johnny Appleseed or John Henry in your project, in your organization, or in your community? One might feel that it would not be possible in every circumstance; the reality is that there is always room for one more myth. Great deeds can always lead to greater deeds and positive role models can also bring about change at many different levels (O'Toole, 1996). As a project manager, one must ask, how can a culturally positive and value-driven myth be communicated in a way that it can benefit the project, while highlighting the deeds or actions of something ordinary or extraordinary? A transformational leader can make the event important, and over time the myth will become larger than life.

In most organizations, the mechanisms to pass along the myth already exist; the mechanisms only need a story to tell. The leader must craft the details of the story to make sure the important elements and the required message are maintained. A transformational leader can achieve this by considering an important milestone about a project, considering how the individual was effective in achieving the desired results, describing the conflict or the challenge that needed to be overcome, and then talking about the story to a few team members or stakeholders. In any social system, the story will be passed along and will eventually also be retold to the individual whom it was about. The myth about bad news traveling fast is accurate, but what the myth does not explain is that good news travels almost as fast. So one can leverage complexity by creating a few myths about people on the team and about the project itself (Rigby, 2009).

Some people would argue that if the story is true, then it cannot be a myth. And spreading positive stories about people is not about complexity so much as rumor mongering (Reddoch, 1990). Yet other mechanistic or transactional leaders would feel that just talking to the person about the achievement would be a more effective way of leadership. Why not just tell them they did a good job, give them an incentive, and then go on to the next task or challenge? A response to all these points—and these points are clearly going to show up somewhere—is that if one cannot reframe leadership style and embrace complexity, that individual will have a hard time trying to cope with social systems that are not as direct as a flowchart would have us realize. Complexity is about letting go some of what we think we know and then using new systems that work. People are social; they are not machines, so if project leaders can look to social systems to help manage, they can be more effective with larger projects and teams that might normally be beyond their reach (Hass, 2009).

Social Myth Example: Chasing Rainbows

When I was a young child, my mother used to tell me a story about rainbows and why they are so important. Everyone knows that sometimes after it rains and the sun comes out, a rainbow might appear. Most of us have also heard about the legend of the leprechaun's gold buried at the end of the rainbow. The legend says that if you find the end of the rainbow, there will be a pot of gold buried where the rainbow touches the ground.

The story goes that one day when a rainbow came out, two entrepreneurs agreed that they would both go out and try to find the end of the rainbow. Try as they might, they did not succeed on their first attempt. So after their failure, they both vowed that they would always chase the rainbow, whenever it came out, in hopes of one day finding the pot of gold at the end of the rainbow. They agreed that whenever a rainbow appeared, they would contact each other and both of them would try to find the pot of gold at the end of the rainbow. Years passed and the two entrepreneurs tried and tried; for years they chased those rainbows. They spent many a day trying to reach the end of the rainbow. They chased rainbows until one day, after many years of being unsuccessful, one of the entrepreneurs went to the other and said, "You know, we have been trying for years to find the end of the rainbow, and I do not think that we will ever reach it. On top of that, I don't think that rainbows ever touch the ground." Having said that, one entrepreneur stopped chasing rainbows. He returned to his work and became quite rich. The other entrepreneur continued chasing rainbows, and to this day he is still trying to reach the end of the rainbow.

And then my mother would ask me, which man was happier?

"Let me guess it."

"OK", she would reply.

"Would it be the man who stopped chasing the rainbow?" I would say.

That is exactly how I would always answer her question, and I would explain in detail that the entrepreneur who stopped was right. Rainbows don't touch the ground and the pot of gold at the end of a rainbow is just a myth. I would explain my position clearly and explain that trying to reach the end of the rainbow was a fool's errand that would never be successful.

Then, after I spent some time explaining my position, she would declare that the man who continued to chase the rainbows would always be happier. She would not explain why, but she would tell me the story over and over again at different points in my life to see if my answer would change.

Needless to say, the story bothered me for years and it took me decades to finally realize the point of the story. I looked at it so many different and logical ways until one day; I changed the context of the story. Chasing rainbows is a metaphor about life. By reframing the story, it made much more sense, at least to me.

What if I told you that a rainbow represented a person's dreams, and the man who stopped was a person who gave up chasing their dreams? If I put it that way, then who do you think would be the happier man? The man who stopped chasing rainbows might be richer, but would he be happier? I kept looking for which person was richer or which person could apply logic to the problem regarding the end of the rainbow. But those are hollow goals.

After all, a dream is just as ethereal as a rainbow. I have found life is much better with dreams than with gold. Perhaps that is why I keep chasing my dreams, regardless of how distant they appear.

The above story is a good personal myth. It raises the individual to a new level and understanding about life. It hurts no one, and it is hard to verify that this story actually ever happened. It is also the kind of emotional story that will be passed along to others, and the story will likely grow in the process. It is this kind of social leadership that makes projects great and makes people feel good about getting involved. Make every project one that can be spun into a myth. Make every project great.

Moving past myths, another easier system to utilize is the metaphor. Humans love metaphors (and similes) because they explain elements of the world into simple and familiar elements. Machines do not care if you connect them to other machines. As long as the systems are connected in a manner that falls within the parameters of the individual machines, everything works—at least most of the time. Human systems operate similarly to mechanical systems most of the time, but now always. Machines do not care about metaphors, but metaphors are important, because, unlike machines, metaphors can help people identify with an organization, metaphors can help people better understand a concept and metaphors can also help motive people past their fears.

An effective transformational leader can take complexity further by inspiring stakeholders to live the social metaphor used to interpret the roles and conditions of the team or organizational success. Transformational leaders can make the workplace experience into experiential-based learning reality for all these connected individuals.

Transformational leaders are people who can envision possible futures and then use that foresight to create a sense of organizational mission. The future can be a metaphor that inspires people to embark on an adventure aimed at realizing a new future of mutual success (Ackoff, 1999). The mission can be inspired by a metaphor, which becomes a touchstone that enables members of the organization to bridge the unknown. For example, suppose there was an organizational project that required the implementation of a new Enterprise Resource Planning system. For most of us who have gone through something like this, a trip to the dentist would be preferable. However, if the transformational leader did some research to understand how the system is like the prior system, might that help in the transition?

One could use a simple metaphor like a growing tree; the tree of knowledge must always grow in order to survive. If one can see that the new system is like the old system, that the new system is just new branches on the tree, might that help with change, rather than having to plant a whole new tree? The new system could then be seen as just a new addition to the already successful family. With some research and supporting information to show how it truly is similar, one can help spread the metaphor to ease the change. The transition will always be painful but one must help it along. In this example, the metaphor is direct, symbolic, and easily drawn back to the project without any difficulty. In other cases, a layered metaphor is necessary.

Some people would argue against the use of a complex metaphor because it is too difficult to understand or they think that some people might not get the message. For the most part, ignore those people—they are all wrong. Complexity of society is immersed in human metaphors. When two people communicate about scenes in a movie they both have seen, do they explain the details of the scenes of the movie? When two people communicate about an event in a book, do they review the details of what happened? In most cases they do not, but there is an innate understanding of what happened that both people can share. They may share how they felt or how it made them feel, yet the details are silent.

Taking this further, how do people communicate about things that are not common to both people? For example, suppose two people wish to communicate about certain scenes in a movie that one person has seen and the other person has not. Some people will try to explain the scene, detail out the participants, and explain the events leading up to the scene, but more commonly people will use metaphors (often from other movies) to explain what happened. People seek to create touchstones or connections in order to describe social interactions. These social touchstones are true representations of complexity. People want to create these touchstones in a way to establish connections. These are social connections and do not follow any pattern. This is at the heart of complexity.

Complexity is about taking something difficult to understand and making it approachable. It is about leveraging chaos, human society, and human relationships in a manner that helps. Metaphors are powerful in projects and can make people connect in ways that make others want to be a part of the project. Following is an example of how a metaphor can be used to guide feelings, values, ideas, and actions.

Social Metaphor Example: Walk the Wall

When I was growing up, my parents and I always lived close to the sea. The ocean had always been a friend to my family. All of my early memories were of the beach near my house. One of the fondest memories of the ocean and the beach was when I was around six years old.

My sister had not been born yet, but I recall my mommy's tummy growing every week. I remember sometimes feeling her belly for the occasional movement of my unborn sister. I recall being puzzled by pregnancy as I had always been told that babies came from hospitals. Nonetheless, Mom insisted on keeping my sister in her belly for what felt like forever.

It was a beautiful early summer day; the air was crisp and the humidity low. I remember the sun slowly setting on the ocean. Mom had let me go down to the park at the waterfront. I remember the miles of sand, the playground, the bike paths, and the pier in the distance. The playground at the time seemed enormous, and often I would spend time there alone in a playground generally devoid of children. It was always pleasant, and rarely crowded. I did not like crowds anyway, so for me it was a secluded paradise.

Along the edge of the playground was a low wall, a barrier between the bike path and the play area. The sand would collect up against the wall in certain parts, making it simple to step up from the playground side to the top of the wall. The wall extended past the playground area, and went on for some time along the bike path. The wall seemed to end when it wanted and extended for quite a long ways.

At the time, I did not realize the playground had been much larger at one time, with many more monkey bars, more swings, a larger play area, and a seesaw. It also had several benches for parents to rest while they watched their children. Years before I was born, a storm had damaged part of the playground and it had been decided in a rather heated community meeting that the playground would not be rebuilt. The damaged items would be removed and a more humble playground would remain. The town council arranged to remove the damaged items, but the wall would remain. I think it was not nostalgia that preserved the wall, but economics. To remove a wall would cost more money, and the wall never hurt anyone, so the town council decided to let it remain.

It was that wall that would be my companion for years. I remember being very young and my mother and father holding my hand to keep me steady as I walked the wall. I learned to balance myself, eventually, and it became a constant game to walk the wall. The wall was tricky because certain parts had decayed and worn away, so in several sections one had to be very careful of one's footing. The sand would also make the top a little slippery, especially, if I moved a little too fast. The wall was like my balance beam, and I was its gymnast.

So, there I was that day of days, walking along the distance of the wall and back. The walk was like my mighty routine in the summer as the sun began to set. While walking along the wall, engrossed in my own Olympic championship delusions, my father suddenly appeared.

It was a little odd, for it was unusual for him to come down to the wall. Usually, Mom collected me if I lingered too long. Also, it seemed a little early for Dad even to be home. He beckoned me toward him since I was a good distance away on the wall. He wanted me to come to him. There was a sense of urgency in his voice. He called my name again, and then suddenly I remembered the days gone by with my father at the wall.

I immediately felt a sense of loss, thinking that he would not walk with me along the wall as he had always done in the past. I cried for him to come, but he remained steadfast in his location. He called again and said that Mommy needed us and that we had to go. I started to weep and called him again to me, sobbing and calling. I told him he needed to walk me along the wall.

I think at that moment something must have clicked in his mind. He must have remembered the moment in time that I had remembered. He sauntered over to me and took one of my hands for balance. We slowly walked the length of the wall in a playful kind of dance. We strolled and giggled. We laughed and sang. At the end, he picked me up off the wall, and lifted me up high, as he always did in the past, and then lowered me to the ground.

Once on the ground, we started back the few blocks to our house. Mom was in the front yard by our car, which was in the driveway. She stood, rather vexed at our behavior. She tapped her foot as we meandered to the car. She said, in clearly an annoyed tone, "I thought I was going to drive myself to the hospital." Daddy apologized and said something about taking too long, but it was worth it to walk that wall one more time. We all quickly bundled into the car and went to the hospital.

The next morning, when I woke up on a couch in the hospital, my baby sister had arrived. I'll never forget that day. But all that was so many years ago now. Now, with my own children, I never miss the opportunity to walk along that wall with them. I still think how much one day and a little walk along a wall can change a person's life.

So, the question is: what do you think about walking the wall? Now that you have read the story, this can become a touchstone for anyone else who reads this book. It becomes a secret code of everyone who reads it. It is a story rich in symbols, communicating values and principles and embracing change. The story affects people differently so there is no reason to try to explain what the impact is. Moving forward, the metaphor of the wall will now have meaning beyond what the wall ever represented. The wall is now an enigma that can be shared with others. It is up to the leader to find a way to translate this kind of metaphor to a project and apply it to the project team in a way to connect everyone together. Making touchstones for people to hold will do more for a project than a thank-you or the individual recognition of a job well done.

Myths and metaphors are elements of complexity that can have far-reaching implications. Some people thank their car in the morning for starting, even though machines do not have feelings and do not respond to myths or metaphors. A great review about a car by Consumer Reports™ does not make an unfeeling automobile feel better, but a positive review in a magazine about a successful project might motive people on that team more than a thank-you from the team leader. The challenge is not whether this will work, but how to make it work. It is hard to understand why this works, but intrinsically it feels like it should work, so the project manager will need to move to manage in this manner in the future. After all, deep inside most people want to be remembered and to be thought of positively by others. Make that dream come true.

GLOBAL CULTURE

Organizations must take into account the impact of globalization upon the organizational culture. No longer can one expect the headquarters of an organization to disseminate the culture of the organization. The more integration that an organization has with international affairs, the greater the influence those situations will have upon the organization (Bolman & Deal, 2003; Schein, 2004). It becomes more important to reflect the organizational culture through individuals and through communication. A project manager needs to make sure that the organizational top performers are delivering results within the parameters of the organizational culture while at the same time making sure the organizational communication is consistent with the organizational culture and values (Bolman & Deal, 2003; Schein, 2004). Failure to do so would allow certain individuals to operate above the culture and would negatively affect the organization. This would further deteriorate the effectiveness of complexity because individuals would then have to choose between following the culture or evading the culture. This will certainly confuse individuals and create greater problems for the project manager as they are forced to address these issues.

In the last 25 years we have seen the rise of manufacturing in Eastern nations, along with the steady decrease of manufacturing in Western nations. This shift in economic growth has come with certain cultural challenges as organizations and nations have come to grips with this new business challenge. There has been a distinct change in the West as cheaper labor and lower environmental standards are making it more attractive to move certain industries to the East. Furthermore, countries like China and Korea seem to be eager to invest in newer technologies to further increase their productivity and reduce their costs long term. This expansion has fueled extensive growth in certain developing nations; there has been a recent cooling in this growth due to new difficulties in the market. However, in order to better handle a cultural challenge such as East and West, one needs to look at certain strategies to combine the strengths of the culture together and avoid unnecessary and costly culture clashes.

The first step in dealing with the clash of cultures is to understand that they are different and each serves a particular purpose. It is imperative that individuals learn to respect another culture before proceeding to the next level. An organization cannot overlook that other cultures will be different and they will have different ideas, morals, values, and beliefs. Some will be driven by the nation, while others will be driving at a more local or individual level. Regardless, one needs to understand and accept the differences in order to move a more harmonious existence. Without respect, there is no trust, without trust there will always be distrust and disharmony. Cultures are not easily changed, so it is critical to start from a place of understanding rather than a position of contempt. If one still feels that the other culture is lesser or wrong, there will be no further progress. One must look past the differences and see what they have in common. In most business situations, there is a strong underlying economic reason to move forward with a cross-cultural relationship. If there is not a strong business case then the two organizations may work together for a while, but it will not yield significant results.

Cultural Acceptance Process

Once acceptance becomes the norm, differences are valued, and commonalities are observed, the cultures can move forward to a more efficient operation. Figure 7.2 shows how each phase builds to the next phase in order to create a lasting bond between cultures. This is a four-step process where culture is slowly accepted and in some cases assimilated by a team or group.

As previously reviewed, the first step in creating a stronger new culture, in some cases a new East-West relationship, begins with cultural acceptance. This is true for any kind of combining of cultures. There has to be acceptance by both cultures in order to move forward. What is interesting about this step is that one side often is typically more accepting of the other side, which can lead to early difficulties as one side perceives itself as trying more while the other side is perceived as trying less. Yet if asked, either group will immediately defend its position as being the group that is trying more. This is a very common defensive behavior that must be overcome early. A project manager must learn to educate all the different cultures of how every group is trying. If an organization observes individuals having difficulties regarding cultures, it is often wise to educate those individual and to have them work with others that are more accepting. Make sure that leaders have a solid commitment toward cultural acceptance and make sure that rogue elements do not develop within an organization.

This kind of education will go a long way in moving forward in the process. As long as there is a perception that one side is trying harder, these groups will never get past the acceptance stage.

> Practical Tip: Part of this concern about cultural acceptance is related to job loss. It is true that some jobs may be lost when organizations and cultures merge, but, theoretically, new jobs should be created as old jobs are lost.

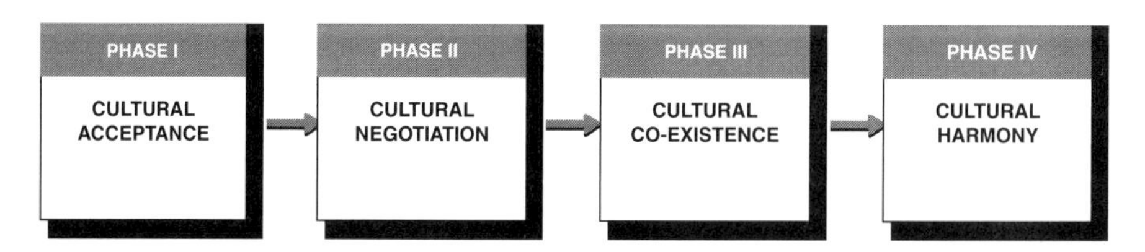

Figure 7.2 Cultural Acceptance Process.

The second step is known as cultural negotiation. This is often where the individuals affected by the cultures will have to negotiate with one another to yield the necessary business or financial results. The importance of a negotiation is that each side will bring certain requirements to the table and they will have to somehow navigate with one another to meet their goals. Also, many cultures place a symbolic significance on negotiation or have certain rituals that need to be performed. These rituals may be conscious of unconscious, depending upon the culture and how the society perceives negotiation. Regardless, one needs to create an acceptable method to complete the negotiation.

If the parties are ignorant of the cultural ramifications of a negotiation, there will be a period of conflict as the parts learn what is necessary. This may cause some discomfort for some individuals, but if there is a cultural acceptance, then each party should actually assist in the process. In other words, if both sides feel that the negotiations are among trusted colleagues, then each side will help the other navigate the cultural requirements to make the negotiation successful. Do not think that the other side will help you reach your negotiation goals; more likely, since one side is less prepared for the cultural aspects of the negotiation, it is likely that a degree of advantage will be taken. That does not mean that a lopsided deal will always result, but it does mean that one needs to be careful about distinguishing between cultural assistance and taking advantage of the negotiation.

For example, if the negotiation requires a level of ritual and if people are mindful and respectful, then each side should recommend a suitable course. If a culture is very hierarchical, then it might require the most senior member to disclose the offer and lead the negotiations. Or it may be considered rude to defer to another authority during a negotiation, when the other side made sure that they have the appropriate decision makers available. One needs to make sure that the proper courtesies are extended.

Two other important aspects of the cultural part of negotiation are to ask what is required and to follow through with what is asked. Most people are happy to offer more information about their culture and to have others respectfully participate. So it is always good to ask, but one must be prepared to follow through with the offer to show that it is genuine. Thus, if it means drinking a token beverage together or shaking hands or possibly sharing a meal, you must be ready to ask and conduct yourself appropriately. Unless you have some very good reason not to participate in different culture requirements, you should learn to respect all the elements of each culture (Rigby, 2009). If there is a good reason, then it is prudent to advise others of the problem and to negotiate an acceptable solution. It is better to meet a culture part of the way with a good reason than to simply ignore the cultural requirement and assume that others will be able to understand the reason by observation.

After the phase of culture negotiation comes cultural coexistence. If all cultures are in equilibrium and there is a common and pervasive trust between the cultures, then cultural coexistence exists. For example, a country that believes in religious tolerance and creates laws and rules to respect all religions is at a point of cultural equilibrium. When all aspects of the culture are protected and respected, the organization has achieved cultural coexistence. This phase is hard to attain and the cultures will have to continue to work toward maintaining this level.

Very few organizations can achieve the final phase of cultural harmony. This is a state where the cultures are so integrated, equal, and trusting that the situation is seamless. Even though few organizations can achieve this level, it is always the state to strive toward; it is possible to have brief periods of success as this level and this type of success goes a long way to keeping everyone satisfied.

Thus, cultural acceptance is a starting place toward cultural harmony. Too often organizations strive to achieve only acceptance as if that were cultural nirvana. It is important to understand that culture issues arise at every level of cultural acceptance; however, as trust, awareness, and respect grow, conflict is less likely to occur. The higher on the cultural acceptance process, the less likely that errors, omissions, and oversights will be perceived as intentional, destructive, or demeaning. These errors, omissions and oversights will then be seen for exactly what they are—terrible yet unintentional mistakes.

> Practical Tip: Consider the difference between feeling that a comment was meant as an insult rather than an inadvertent mistake. There is a lot more intent in an insult, which an individual takes personally and sees as less forgivable. By contrast, an inadvertent mistake is considered more forgivable.

CHAPTER SUMMARY

Culture is what guides a group toward certain social norms and accepted behavior. In some cases culture can be reinforced by certain standards, such as a professional dress code, high work expectations, and other organizational frameworks. Culture helps define the high-level organizational expectations that help maintain focus for an organization.

Culture can impact leadership; good leadership will mold itself around the leader's style in a way to keep individuals focused. Any leadership style can be successful as long as it incorporates the different elements of the culture. A successful culture and a successful leader is when the two are blended together in order to achieve a level of transparency. The leader must embody the culture in order to magnify the effectiveness of the team.

Quality is also important in organizations because a culture of quality can pay dividends in high work output and fewer defects. Every organization has a level of acceptable defects, but when the culture can help reduce this amount, it becomes even more successful than the competition. Organizations might not achieve six sigma levels of defect-free results, but the organization might reduce defects to a point where it can become a significant organizational feature.

Globalization has impacted organizations by leaps and bounds. No entity in any nation can claim that it is not impacted by the international markets. The global nature of energy, food supply, industrial goods, and even knowledge has created a global village where everyone is the competition. The more organizations can leverage this aspect of life, the better off they will be. Furthermore, an organization—and more importantly, the leaders of any organization—must recognize and lead according to what is needed in the global market, not what is needed by the domestic market.

Thus, organizational culture is important, but keep in mind that it is only a subset of the greater global culture. Organizations will succeed or fail in the future based upon their ability to harness their organizational culture and how well they can blend this into the culture of the global village.

REFERENCES

Ackoff, R. L. (1999). Transformational leadership. *Strategy and Leadership*, *27*(1), 20–25.

Bolman, L. G. & Deal, T. E. (2003). *Reframing organizations: Artistry, choice and leadership*. (3rd ed.). San Francisco: Jossey-Bass.

Carin, A. (1997). *Teaching science through discovery* (8th ed.). Upper Saddle River, NJ: Prentice Hall.

Cooke-Davies, T., Cicmil, S., Crawford, L., & Richardson, K. (2007). We're not in Kansas anymore, Toto: Mapping the strange landscape of complexity theory, and its relationship to project management. *Project Management Journal*, *38*(2), 50–61.

Duarte, D., & Snyder, N. (2006). *Mastering virtual teams: Strategies, tools, and techniques that succeed* (3rd ed.). San Francisco: Jossey-Bass.

Hass, K. (2009). *Managing complex projects: A new model*. Vienna, VA: Management Concepts.

Krajewski, L. & Ritzman, L., (1996). *Operations management, strategy and analysis* (4th ed.). Reading, MA: Addison-Wesley Publishing Company.

Lipnack. J. & Stamps, J. (2000). *Virtual teams: People working across boundaries with technology* (2nd ed.). New York: John Wiley & Sons

Marcinko, R. & Weisman, J. (1997). Leadership secrets of the rogue warrior: A commando's guide to success. New York: Simon & Schuster.

Osborne, M. (1991). *American tall tales*. New York: Alfred A. Knopf.

O'Toole, J. (1996). *Leading change: The argument for value based leadership*. New York: Jossey-Bass.

Project Management Institute (Ed.). (2008). *A Guide to the Project Management Body of Knowledge—Fourth Edition*. Newtown Square, PA: PMI.

Reddoch, J. (1990). *Life shall be what it shall be*. Huntsville, AL: Huntsville International Press.

Rigby, D. (2009). *Winning in turbulence*. Boston: Harvard Business Press.

Schein, E. H. (1992). *Organizational culture and leadership* (2nd ed.). San Francisco: Jossey-Bass.

Singh, H., & Singh, A. (2002). Principles of complexity and chaos theory in project execution: A new approach to management. *Cost Engineering*, *44*(12), 23–33.

Chapter 8

Cultural Conflict through the Lens of Complexity

Figure 8.1 The warning sign offers a potential lens to a possible future. Most project managers will not encounter a mountain lion along a lonely road, but the sign gives a metaphoric glimpse of what dangers might lay ahead. Conflict is a similar problem that can arise without warning, so a project manager must remain prepared even for the improbable.

TRADITIONAL CONFLICT

Conflict can happen in any team at any time, so the prudent PM should anticipate conflict within his or her teams. Sometimes conflict can be constructive, such as the conflict that comes when groups establish their roles within the team. Conflict can also be destructive, such as when team members do not cooperate and actively undermine one another's efforts (Duarte & Snyder, 2006). Destructive conflict can be more prevalent in virtual teams as team members must overcome communication issues due to technological constraints in addition to having to work with issues of trust (Jarvenpaa, Shaw, & Staples, 2004). Excessive internal competition within a virtual team tends to erode trust and increases the possibility for conflict. Team members must overcome communication issues due to technological constraints.

Another of the many challenges within the virtual community is the conflict resolution and the skills necessary for the environment. Individual companies may assume that conflict may be the same in the virtual environment, and the skills necessary are similar to those of a traditional brick-and-mortar company (Tuckman, 1995). By assuming that the skills necessary are the same as traditional teams, the PM may inadvertently make some critical mistakes.

The PM must have an understanding of the critical differences between teams and virtual teams. A traditional team communicates directly through face-to-face contact (Tuckman, 1995). This direct communication includes body language, tone, and other visual cues, which are important elements of communication (O'Conner, 2000). Virtual teams regularly contend with the lack of direct contact and rely on indirect communication, in the form of telephone calls, e-mails, faxes, and other technologically based methods.

The virtual environment requires that communication be clearer and more concise. Technology cannot replace poor communication; in fact, effective communication is one of the most critical elements of a virtual team (Duarte & Snyder, 2006; Roberts, Kossek, & Ozeki 1998). Furthermore, the leaders of the organization must support the virtual team by providing the necessary technological resources for these types of projects. To overcome the challenge of communication in the virtual environment, many studies and documented practical experiences emphasize the need for trust in the virtual environment.

Practical Tip: Conflict will reduce efficiencies and sap the strength of any team. The leader must remain cognizant of any friction among team members. Awareness is the first step in addressing conflict.

VIRTUAL CONFLICT

Many trained and experienced project managers are versed in the areas of conflict and conflict resolutions for traditional face-to-face projects. However, there has been little documentation regarding effective methods to resolve conflict on a virtual team. Since the virtual PM has minimal verbal and nonverbal clues to pending conflict, a PM must learn to address and mitigate conflict in a less robust environment (Duarte & Snyder, 2006). Conflict will always happen, but a virtual project manager must learn how to address and mitigate conflict in difficult circumstances.

First, the PM must learn to address conflict. The PM must learn that the virtual environment requires a more direct approach to conflict. Since there are fewer nonverbal cues, the PM must learn to directly deal with the conflict.

Addressing conflict is an important function of the PM. The PM must learn to address and resolve conflict because conflict will cost the project time and money. Addressing conflict has two stages. There is the discussion or communication stage, where the PM tries to review and discuss the crux of the conflict. Often listening and reiterating the problem with the affected parties can assist in explaining what the source of the conflict is. Often there is more miscommunication than actual conflict; however, it is important to try to find out the source. Reviewing the problem can help develop a solution. Too often in the virtual environment, the affected parties avoid contact in order to avoid the problem. This method of avoidance is often detrimental to the project since it causes the involved parties to miss out on important communication. The PM must resolve the issue in order to get the project back to functioning effectively.

Second, a PM must learn to mitigate conflict. Speed is the best way to mitigate conflict. The sooner the PM can identify conflict and address it, the sooner the project can resume smooth operations. The worst thing for any conflict is for it to linger. The longer it lingers, the more it festers, and the more it festers the more it infects. Conflict is like a disease that slowly infects and destroys a project (Johnson & Johnson, 2000, Tuckman, 1995).

Furthermore, a PM must learn to identify two distinct types of virtual conflict: personality conflict and task conflict. Both of these types of conflict are important to understand, and distinguishing between the two will assist in the correct approach. The PM must also understand that many times both of these types of conflict are mingled together, particularly if the conflict has been simmering for a long time. Both these types of conflict require communication and clarification; however, task conflict requires clearer responsibility definitions.

Personality conflict is when two or more individuals do not get along due to some stylistic way of handling communication or other people. Many times people engage in this type of conflict via a third party. Sometimes people will hear or be involved indirectly with the conflict, which often makes the conflict more difficult to control.

> Practical Tip: Virtual conflict is often harder to find than normal conflict. Conflict is easy to spot in a face-to-face situation because certain individuals will not work together well. The best approach for the leader to take with conflict is to ask directly and periodically. Most people want to talk with someone else about a problem they are having, so taking advantage of this fact can serve to help better address virtual conflict.

POSITIVE CONFLICT: THE ART OF NEGOTIATION

Many times conflict happens among team members because they are negotiating for a particular position or item. This kind of conflict is not negative, but is only part of the negotiation process. All things of value, by definition, are points of negotiation. All individuals must negotiate for all things. The problem is that sometimes when people are negotiating for a position, for an item, for power, for the interests of their organization, for their own interests, there is often some form of friction between individuals. Negotiation is one of the most important skills in the free market society, yet few institutions place any significance on this skill. The reality is that the art of negotiation is perceived as a skill and nothing more.

Negotiation happens every day. In every negotiation there are four important stages that should be followed. Each stage must be addressed if one is to maximize the output of each negotiation. These stages are research, plan, action, and review. All of these activities are important for negotiation; one must understand that conflict occurs during the action phase when an offer is made.

Research

The research phase is the most important phase of any negotiation. One should never enter into a negotiation without understanding the position and expectations of the other side. The research phase consists of three different fact-finding areas: understanding the position of the other side, understanding the capabilities of the other side, and understanding the retribution of the other side.

Understanding the other side is always the first fact-finding area of any negotiation. One must always ask and establish the expectations of the other party. To negotiate properly and to achieve a positive outcome, one must understand the needs of the other side. Often conflict occurs because one does not truly understand the needs of the other side, so when you negotiation with them, they react negatively to the offer. It is important to understand the organization and the people you are negotiating with; however, it is more important to understand their position. Much as one would not vote for a political candidate without first understanding their platform, one should not negotiate with a party without understanding their platform. Establish the exact criteria of the other side's desired outcome. Understanding their position and desired outcome from the negotiation is critical to making the conflict positive rather than negative.

Finding out this valuable information is often quite easy. The first method to finding this out is to ask. Typically in business negotiations, most individuals will reveal most of their objectives up front to anyone who asks. Do not take these statements in a vacuum. There will often be additional items on a *hidden agenda*. For example, when buying a house, the seller will tell you their price, but that price is often the highest price they would like to receive. They usually do not disclose the lowest price that they will accept. Often the other side will let you know their short-term needs and goals, but will hide their long-term strategy. This is usually due to the fact the individual or company feels that this type of information is sacred. Interestingly, this type of information would be most helpful if one is truly to find a long-term, mutually beneficial agreement. How can one make a lasting long-term agreement if one does not understand the long-term needs of the other side? This may hinder the negotiations up front, but there are methods to find out this information about the other side.

The second method to find out this information is to research the company or the person in question. There are several ways to do this. One can examine what information is posted on the Internet. This readily available source of information will have tons of information about almost every company and often many people. Furthermore, one can also research company information in a library. A quick search of the company name can yield any number of recent documentation. One can also examine information posted about publicly traded companies, through 10-Q or annual reports. The growing trend has been to include much more than the required financial reports about a company. This may not give you anything more than statistical and financial information about a company, but press releases or other associated documents or reports can often give valuable information about a company. At a minimum, review what the company is advertising to stockholders. This kind of information will give valuable insights about what the other side will be willing to offer in a negotiation.

The third method is to have a third party research the company. This can be done by getting a Dun and Bradstreet™ or some similar report on a company. If possible, have your bank make the inquiry, because they will have a more extensive report than what is typically offered to businesses looking for credit reports. This type of intense research is probably necessary only if one were looking for partnerships or joint ventures because it can be costly and timely. Background checks early on can save valuable time and embarrassment when negotiating close interactions between companies.

Plan

When a point is reached that one feels all the appropriate information about a company has been researched, it is time to formulate a plan for the negotiation. A plan should have a definitive beginning, middle, and end of the negotiation. The negotiation plan should establish what kind of opening arguments or discussion should take place to firmly explain your position to the other side.

Once it is determined what needs to be communicated in the beginning, you then need to formulate what is said in the middle. The middle section should have two important aspects. First, one needs to be able to firmly communicate one's position in the negotiation. Make a list of all the needs, wants, and nice-to-haves of the negotiation. Second, formulate contingencies based on how the other side might react to the points presented or what they might be expected to ask for.

The first part of the middle section of the negotiation is a period ripe for conflict because the other side may not agree with your position. Hopefully, you have made a convincing argument in the opening to sway your opponent toward your position. This is the point in the negotiation when there is a lot of give and take. In the plan, one needs to map out the needs, wants, and nice-to-have items and be prepared to dole out concessions. A good plan maps out what one might be willing to trade for something else. For example, if Y is being offered to one as a concession, then in turn X should be offered in trade. Maintain the trading so that the actions remain positive. Sometimes conflict arises when the other side feels that the situation is not fair or when one side is taking advantage of some situation. If one gets a sense of that from the other side, it is time to back off and approach from a different direction. This is when one needs to have some good contingencies.

The second part of the middle section of a negotiation is all about contingency. One needs to plan in advance what to offer or do when the other side takes a certain position. For example, if the other side acts belligerent in hopes of intimidating the opponent, one might have a contingency to walk away from the negotiation. One might propose a cooling-off period so the other side will have time to think about their actions. Be prepared for different actions by the other side; often writing a list of what may happen in the negotiation is a good way to cover all the possibilities.

One important contingency to consider is when to walk away. No deal *must* be done. If you establish in advance what the limit is, then you need to stick to that limit, no matter how painful it may seem. For example, if you have determined in advance that you can afford a house that is $200,000 and the other side is stuck asking for $225,000, then you must walk away from the deal. It may seem hard, and a difference of $25,000 might not seem like a lot of money over the period of a mortgage, but if the limit is $200,000, then stick to the limit.

The end portion of the middle phase is all about the closing. Regardless of the outcome, one needs to make sure that the negotiation ends on a positive note. It may seem strange to want to keep it positive, but consider that ending on a positive note enables both sides to walk away from the negotiation feeling victorious. It is better for people to part feeling good about the deal than to have one side come back and try to renegotiate the terms again.

> Practical Tip: No deal is so important that you cannot live without it. The moment you feel that you must make the deal is the moment that you have given up your walk-away power, and it is also where the other side can take advantage of you. When a person is unwilling to walk away from a deal, they will always end up with a less than equitable deal.

Action

The action portion of a negotiation is the interaction between the two parties. Again, this is when conflict might happen. People that passionate are bound not to agree with someone who is passionate about an opposite position. One should have some plans on how to handle conflict that might arise. One should also make sure that the conflict does not interrupt the negotiation. Sometimes if there is a break in the flow it will slow down the process and it might also make people more resistant. If conflict does arise, it must be addressed during the negotiation. It needs to be corrected prior to moving on because conflict will only serve to slow down the process or to make people less likely to deal at any cost.

As discussed in the plan, one needs to make sure to end the negotiation in a positive manner. The action should close on a positive note where everyone leaves the negotiation feeling good about the deal (or lack of a deal). The close should be one in which everyone feels the best possible situation has been reached. Too often people feel that something was left on the table, and that causes feelings of regret and remorse. This is not a positive way to end a negotiation, so make sure to close in a way that is amicable.

Finally, the action phase should end with a thank you to the other side and a compliment on their negotiation skills. Thank the other side for making the deal (or not making the deal) work for both sides. Find a particular point where the other side did a good job at negotiation and compliment them for that point. Make sure to recognize the positive and businesslike attitude of the negotiation and that both sides did not take the problem personally.

Review

This is an important section of negotiation that is often overlooked. Once a negotiation is concluded, review in your mind what could have been done better. Also review what should be done in a similar case in the future. It is not easy to critique one's own negotiation style, but it is good to revisit it for improvement. Also consider garnering feedback from others on the negotiation team (if there was a team) about what went well and what could be improved. In particular, if there was conflict, review how it was handled and consider whether that was the best way to handle the situation. Often conflict is difficult to deal with, so objectively reviewing it afterward will help give a better perspective on how to handle a similar situation in the future. There is always room for improvement, so the better you can learn from the past, the more equipped you will be for the future.

NEGATIVE CONFLICT

Negative conflict occurs anytime team members are not operating cooperatively toward a common goal, due to past external personal or professional issues that affect their current disposition toward a task. Negative conflict results in both wasted time and lost productivity for a project because conflict costs the organization money and is a contributing cause of project delays. It is estimated that the cost of conflict, when including ineffective managing of interpersonal situations, conflict avoidance, and lost project days, accounted for $20,000 per employee, per year, in addition to 20–25 percent of a manager's time spent dealing with team disagreements (Johnson & Johnson, 2000, p. 337). The virtual project manager must identify potential periods of conflict as well as understand strategies to cope with destructive conflict, also known as subversive conflict.

> Practical Tip: Sometimes it is difficult to recognize negative conflict. If you are ever unsure, ask impartial stakeholders to review the information. If they feel it is a negative conflict, meet with the individuals involved and ask them. Sometimes by saying that an external party views it as a negative conflict, this can diffuse some of the defensive behaviors that team members might have if you were to try to directly approach a destructive conflict situation.

ADDRESSING SUBVERSIVE CONFLICT WITH COMPLEXITY

Feedback has always been perceived with some resistance due to some negative experience many of us have had while receiving painful feedback. There is a known relationship between decisions and feedback, and decisions and negative feedback. There are two types of subversive conflict, the conflict when there is a lack of consensus and conflict that occurs when an individual feels disenfranchised by a decision. One would naturally assume that the workplace would be more positive, but the reality is just the opposite.

Consensus is more important in areas or places with a low level of loyalty to the workplace. People desire more consensus when leadership or direction of the company is unclear. There is a definite correlation between organizations that might have major changes within a few years and the need for consensus in decisions. For example, if a company might be sold in the near future, the employees need to be assured that management is listening to them. This means a level of trust supported by consensus. Without this, or when there is dissent in decision making, levels of morale drop significantly. When morale drops there is more subversive conflict within an organization. Subversive conflict consists of individuals who conflict with the organization in hopes of maintaining the status quo.

This type of subversive conflict can be addressed by making sure that consensus is reached and communicated. This ensures that individuals feel part of the process and feel that they have input into the decision making process (Townsend & Demarie, 1998).

In some cases, when a decision is reached through a generally accepted method and one person does not agree, that person takes on the role of the dissident. This type of conflict, which occurs more in areas where loyalty and morale is high, is seen as disruptive because it does not help the organization move past certain decisions.

This kind of subservice conflict can be avoided by allowing appropriate forums (Rigby, 2009). Even if decisions are taken with which the majority does not agree, they will feel that they had their opportunity. To avoid subversive conflict it is important that people are heard, but it is not as important for all their decisions to be reflected within management. People understand that all their needs cannot always be addressed in a project, but if they feel that most are being listened to they are more supportive of the organization as a whole (Hass, 2009). Thus, conflict cannot be avoided in different environments, yet most people are at least satisfied when majority opinion is considered and often implemented.

> Practical Tip: When subservice conflict arises, bring it out into the open. If you bring it to the attention of others, they are less able to subvert the team. If others know of their intentions, they are less likely to be influenced by them. If you approach them directly in a public forum they have a choice to deny or refute the claim. If they deny, then they are less likely to be able to influence others in the future. If they agree, then it is sufficient grounds to remove them from the project.

IDENTIFYING AND CORRECTING LACK OF COMMITMENT BY TEAM MEMBERS

In the virtual environment, the project manager does not have the opportunity to manage by walking around, to have face-to-face discussions with members who are having issues, and to diffuse conflicts with all members present. This presents challenges for the virtual project manager. There are several manners by which the virtual project manager may identify those who lack commitment. Then the virtual project manager needs to address team members who are not meeting expectations.

The virtual project manager must make him or herself *visible* within the team, and hold the team leaders and team members accountable. Leadership of the virtual team needs to establish high expectations for team members to ensure success and to hold the team leaders and team members accountable to these expectations. This, in turn, creates trust. Trust is an integral aspect of a virtual team, and studies indicate that without this trust a virtual team will more likely fail. Trust enhances communication, which is essential in a virtual environment and, in fact, alternate channels of communications should be encouraged.

A team member who is not performing may not be suited for the virtual environment. Some individuals need daily human contact. This should be assessed prior to the virtual project assignment, if possible. However, if it is not, the virtual project manager should discuss the expectations of the virtual environment with the team member and leader. Remember to document all performance-related discussions with any employee.

Should the team member not be suited for the virtual environment, remember that you are responsible for the project. While it is not pleasant, you may have to ask the team member to leave the project. Remember to follow the policies and procedures of your company and involve the human resource department.

Remember, in the virtual environment it is essential that each team member understand his or her responsibility and that he or she will be held accountable. You, as the virtual project manager, *must* ensure this is done, is communicated, and is visible. In this situation, the virtual project manager should be able to assess the performance of the team member. Counseling should take place for all team members on a regular basis or as required by the company's policy. As before, follow your company's policies.

> Practical Tip: Accountability is very important, as is understanding the policies of the organization. Many times it is necessary to offer feedback, both negative and positive, in order to focus on the stated objectives. Whenever offering negative feedback, make sure to stay within company policies and procedures, because being too harsh or too soft will certainly affect the feedback.

Team members regularly come and go from a project for various reasons. The responsible project manager should implement roll-on and roll-off procedures during the planning phase of the project. Some extra steps may need to be added since the project is in a virtual environment. For example, do the laptops need to be sent to a central location when a person leaves; does the resource manager need to be notified; and where does the person's project collateral need to be placed?

These are just some items that need to be addressed. The virtual project manager also needs to think about an exit interview and lessons learned. The exit interview is impersonalized in the virtual environment, as it may have to be done via the phone. However, the virtual project manager or team leader should make every attempt to schedule a face-to-face exit interview, if possible, or at least a videoconference.

During this process, the virtual project manager should have an established exit interview, depending on the situation. Is the individual leaving the company or is the individual leaving the project for another project within the company? These are two entirely different exit interviews. In both situations, the virtual project manager wants to establish lessons learned for the project, for the organization, for the company, and for any external risks that could be handled differently.

As individuals leave the project, the virtual project manager will accumulate data on lessons learned. The intuitive and *smart* virtual project manager will have the project management office analyze the results and look for trend analysis. He or she will ensure that low-hanging fruit is implemented quickly and efficiently. Those lessons learned that are outside the realm of the project should be quickly and succinctly sent to the steering committee, the organization's and corporation's leadership.

> Practical Tip: It is often wise to carefully consider the feedback in any exit interview. Many times the most passionate feedback that a person can give will be during this process, so it is often prudent to consider what information is being offered in this process.

CHAPTER SUMMARY

Conflict is more common in organizations that focus on teams than in organizations that seek to isolate individuals. In the past, the strategy of management was to isolate individuals in order to prevent them from organizing against capitalistic organizations. As trade unions and other organizations grew in the industrial age, shrewd capitalists saw the power and energy of organizations. They understood that if these organizations could be focused toward the goals of the organization, teams could ultimately become a highly productive force. Unfortunately, all teams do not operate together at all times. Many times conflict can reduce the productivity of a team and in some cases can bring down an entire organization.

Conflict can be highly destructive when team members actively undermine the team's efforts. Internal competition within a team, virtual teams included, tends to erode trust and increased conflict within teams. Teams must learn to reduce the amount of negative conflict, while still preserving the opportunity of differing opinions within the team. The resolution of conflict is one of the many challenges within the virtual community. Companies may assume that conflict is the same in the virtual environment as in traditional brick-and-mortar company, but virtual conflict is often less apparent. The PM must have an understanding of the critical differences between teams and virtual teams. A traditional team communicates directly through face-to-face contact. This direct communication includes body language, tone, eye contact, and other visual cues. Virtual teams regularly contend with the lack of direct contact and rely upon indirect communication, in the form of telephone calls, e-mails, faxes, video conferencing, and other technologically based methods.

The virtual environment requires communication to be clearer and more concise. Roles and procedures need to be clearer and less ambiguous, while remaining flexible to accommodate future change. Technology cannot replace poor communication; in fact, effective communication is one of the most critical elements of a successful virtual team. Trust is considered one of the critically important factors in virtual teams. Trust becomes important even when team members are released; when people trust the organization, the exit interview process can often lead to some important findings about the project and the project process.

CASE STUDY

Innovative design equals success. In the fast-paced world of work, companies can no longer utilize traditional methods to maintain a competitive advantage in any marketplace. The successful companies of the future will create a design advantage through the strategic leveraging of experts on a short-term basis. Exceptional design will be the bailiwick of small entrepreneurial groups with a talent for surprise. For those who care to disagree with these bold statements, one only has to look at the recent success in the technology sector. Design advantage is understood in the software industry. Small software development companies painfully create innovative products and ideas that surpass the creations of large software companies. Slowly these larger companies adopt and adapt what these smaller companies have done and integrate these designs into their software. The large software companies continue to thrive; the smaller companies are the true design innovators. It is these smaller companies that will always offer creative solutions well ahead of the competition. Nowadays, the design advantage is not limited to the technology industry. All industries are taking notice of the seriousness of the design advantage. Companies must learn how to maximize the innovation potential in their sector, or else they will slowly succumb to the competition. Project Management LLC (PML) is a company created to exploit the design advantage in the area of commercial property development.

Project Management LLC (PML) was established in June 1998 as an independent development organization specifically formed to meet the needs of property owners. PML is fully equipped to perform the services of a development outsourcing firm. In the past, development services have been kept internal to a shipping company. Product development has traditionally been a company's closely guarded secret. The reality is that most companies are aware of what a customer desires; however, they are simply incapable of leveraging the appropriate resources for the desired results. Sometimes a company is so burdened with internal red tape that any innovation is stifled in the internal design process.

PML has the ability to create value by handling all the subcontracting work that owners in the past have had to handle. By offering services that make the project easier for owners, they are better able to work with the investors to better meet their needs. In addition, the owners can then spend time with the property managers who will handle the project after the development phase. A project team can be designed to meet the unique needs of any operator. The purpose is to create new value from services that have traditionally been costs centers within an organization.

PML has grown by leaps and bounds since its inception; the problem is that this company is now being challenged internally. As the organization has had to hire more and more new people to meet the demand, certain individuals have joined the company with their own individual agendas. Some people have also had their own ideas about expanding their departments and have not been concerned about the goals of the organization. Two such individuals are Sam Lee and Suzie Johnson. Both of these individuals have had their own agendas since joining PML and they are now beginning to affect the organization as whole.

Sam Lee has been a productive and responsible project manager but now he has begun to speak poorly about the company and has been very negative. His negative attitude has affected his team and many people are looking to leave his team as soon as possible. His productivity has dropped and he has told certain individuals privately that he is looking for a different job. He is clearly being subversive in his conflict as he attempts to undermine the organization.

Suzie Johnson has been different with her conflict as she has lost focus and is not as productive as before. She is in a virtual team and has been a stellar performer in the past, but now she seems to take things less seriously and is not as focused. Some people claim that her personal life is not as good as before, but she has suffered other misfortunes in her life before and it never affected her productivity. She recently came into conflict with another team member as the team was distributing work for a recent task and Suzie complained that she was being given too much work. She has not been a complainer in the past, but she exchanged some unfortunate words with some of the other team members that have left the team in a precarious state.

- How would you address Sam Lee's subversive conflict?
- How would you negotiate with Sam Lee so that he can gain focus in order to become a productive project leader?
- Would you consider releasing Sam Lee from the organization? Why or why not?
- What do you feel is the problem with Sam and how might you address it if you were his boss or his subordinate?
- Given the background of the organization, how might you think that Sam Lee's conflict is affecting the organization?
- How would you address Suzie Johnson's conflict?
- How would you negotiate with Suzie Johnson to regain focus in order to become a productive team member again?
- Would you consider releasing Suzie Johnson from the organization? Why or why not?
- What do you feel is the problem with Suzie and how might you address it, if you were her boss or her subordinate?
- Given the background of the organization, how might you think that Suzie Johnson's conflict is affecting the organization?

CASE STUDY REVIEW

This is a classic case of the loss of organizational trust. The organization and individuals are losing their trust in the company and in those around them. The individuals are starting to see only the negative and need to have something more constructive to consider. Many individuals would see this opportunity as one to start "chopping out the dead wood." Yet the individuals in questions are proven performers. There is good reason for greater concern as these individuals have some internal clout within the organization.

Their immediate supervisors must take steps to bring them back to understanding the organization. Steps must be taken to address this kind of subversive and indirect conflict. Although this might not been seen as conflict, it has all the elements of conflict and must be handled accordingly. The sooner the action is taken, the better and more concentrated the response. Consider that these individuals are proven performers who hold a lot of project knowledge with them. Their loss would impact the organization in the future. These are the kind of individuals who cannot be easily replaced, so it would be a better use of resources to address the conflict and move past this situation than to try to replace them.

> Practical Tip: Consider the replacement costs of top employees and compare that to the cost of addressing conflict. It is far more economical to keep a good team together than to try to build a new one whenever moral drops.

REFERENCES

Duarte, D., & Snyder, N. (2006). *Mastering virtual teams: Strategies, tools, and techniques that succeed* (3rd ed.). San Francisco: Jossey-Bass.

Hass, K. (2009). *Managing complex projects: A new model.* Vienna, VA: Management Concepts.

Jarvenpaa, S., Shaw, T., & Staples, S. (2004). Toward contextualized theories of trust: The role of trust in global virtual teams. *Information Systems Research*, *15*(3), 250.

Johnson, D. & Johnson, F. (2000). *Joining together: Group theory and group skills*, Needham Heights, MA: A Pearson Education Company.

O'Connor, C. (2000, August). Building the virtual team. *Accountancy Ireland*, *32*, 20–21.

Rigby, D. (2009). Winning in turbulence. Boston: Harvard Business Press.

Roberts, K., Kossek, E., & Ozeki, C. (1998). Managing the global workforce: Challenges and strategies. *Academy of Management Executive, 12*(4), 93.

Townsend, A., & Demarie, S. (1998). Keys to effective virtual global teams. *Academy of Management Executive*, *12*(3), 17.

Tuckman, B. W. (1995). Developmental sequence in small groups. In T. Wren (Ed.): The leader's companion: Insights on leadership through the ages, pp. 355–359. New York. Free Press.

Chapter 9

Cultural Conflict Resolution Strategies

Figure 9.1 The fearsome Tyrannosaurus rex is known as the king of the dinosaurs with good reason. Its size and its carnivorous appetite made it a fearsome opponent to any land creature. Culture conflict is no different; it can destroy an organization faster than most other threats in the world. A project manager must learn to respect the culture of all organizations encountered while still maintaining order. If conflict does arise, the project manager must learn to deal with this threat quickly and efficiently or suffer a tragic fate at the hands of a cultural T-rex.

INTRODUCTION TO CONFLICT RESOLUTION STRATEGIES THAT DEPLOY COMPLEXITY

The virtual team deals with temporal issues. Most virtual teams deal with asynchronous environment and use various techniques to stay current with the other team members, including synchronizing e-mail/databases and conducting meetings at periodic intervals (Duarte & Snyder, 2006). This may lead to confusion and tension within the team, which in turn leads to conflict. Virtual relationships are some of the most difficult relationships in which to develop trust. Trust often develops slowly and requires considerable reinforcement to reach the same level as relationships between proximate individuals. Therefore, the virtual team will have difficulties in resolving team conflict.

> Practical Tip: There are two types of conflict: healthy conflict and destructive conflict. When trust is prominent on the project, team members feel comfortable expressing dissenting or different thoughts about resolving issues on the project. These dissenting options can lead to some conflict, and could even appear to be heated, but since trust is there, all involved know that their opinions will be valued. On the other hand, when trust is lacking, heated conflicts will quickly spiral out of control, team members will not speak, and some may even actively sabotage one another.

The project manager has to be able to develop trust from the onset of the project or the forming stage and should demonstrate his or her competencies in project management (Duarte & Snyder, 2006; Tuckman, 1995). The project manager should be direct and provide an authoritative persona at this stage. He or she should ensure the team has been introduced and understands the project charter and project mission, and should ensure that up-to-date project documents are available online. The project manager should also often include the sponsor and the senior management. This will encourage trust in the project manager and management. During this phase, it is also beneficial to have face-to-face meetings whenever possible.

> Practical Tip: How can a project manager create trust in a virtual team? Suggest that each team member have a way to visually communicate, either by PC camera or videoconferencing so that a face can be placed

with a name. Another way to build trust is to call the team members on a regular basis, just to see how they are. Show team members you've remembered birthdays and family events by sending an e-card or calling them.

USING COMPLEXITY TO PREVENT CONTINUOUS LEADERSHIP INTERVENTION

The project manager should confront all conflict quickly. If it occurs early on in the team formation, quell this dissent early in order to achieve team cohesiveness. Once the conflict is resolved, the project manager should slowly shift the leadership to encouragement of team building and holding individuals accountable for deadlines and deliverables (Curlee & Gordon, 2002). The project manager should be aware that he or she might have to remove disruptive individuals from the team. As the project progresses, the project manager should shift the leadership to focus on removing barriers that the virtual team may encounter at the home office or with bureaucracy on the team.

Practical Tip: Handle conflict quickly and efficiently. Address the problem immediately and openly with those concerned. Try to rebuild trust by discussing the problems with those involved and other members of the team.

In the case of a conflict intervention, the project manager should attempt to conduct the intervention face to face or by phone. Never try to resolve conflict via e-mail. These situations are always difficult and the more robust the communication mode, the better. Also, it is important to follow up after the resolution to make sure the situation has been resolved to everyone's satisfaction. A virtual team leader must keep in mind that individuals from different cultures might have a different perspective on conflict and conflict resolution.

Virtual team leaders must also learn to communicate more effectively with their team members (Cascio, 2000). Group leaders should consider adjusting their leadership style in order to meet the developmental requirements of the group. Effective leadership is often found to be a key predictor of business success, and ineffective leadership is a key predictor of organizational failure. Since leadership is so important to organizational success, it seems only natural that organizations should strive to understand the role of leadership and the role of the leadership style within an organization. By leveraging this better understanding of team members, a leader can learn to better communicate with various team members at different stages of the project.

> Practical Tip: Communication is best when repeated and understood by all. When an important discussion takes place, make sure to document the meeting. When an important communication is issued, make sure to keep a copy and pass along the communication to all the stakeholders and offer a follow-up on the communication at a later point to emphasis the issues addressed.

PREDICT CONFLICT EFFECTIVELY

Understanding the development process of a virtual team can aid in knowing when conflict will occur. Tuckman's model (1995) defined five phases of team development: forming, norming, storming, performing, and adjourning. This model of group development has applicability to virtual teams. The virtual project manager faces conflict during the forming and storming stages of the Tuckman model, which was the same as the brick-and-mortar project manager. Research has shown that trust is essential for effective management in the virtual environment. Project managers must be competent by effectively communicating with customers and teams, and setting high expectations. It can be reasonably concluded that a competent project manager must gain the trust of team members through effective communication.

The forming and storming phases are the two most likely times when internal group conflict might occur. Since the forming and storming stages are the first two phases of a project team's life, the team has not been able to establish the necessary trust with the project manager. It may be effective for the project manager to meet face to face, electronically, or traditionally, and demonstrate competence as a project manager to increase the trust quotient. In these phases, a project manager must establish a clear group hierarchy to avoid conflict. These roles shape a person's future contribution to the group. A project manager who is working virtually should expect conflict during the team formation process.

Communication limitations exist in the virtual environment, so the project manager must plan to compensate for this initial formation delay. Teams start to form and individuals must learn to interact in the new environment. The distance and possible cultural obstacles can retard the initial team formation. Individuals unfamiliar with one another will take time to get to know one another. Usually this kind of formation occurs naturally through contact within an office. Without this kind of interaction, virtual teams may form slowly as people become more familiar with other team members. The formation of small teams within the team helps the overall virtual team become more successful. This allows for the creation of smaller, more agile teams.

A project manager can avoid conflict during the forming stage by increasing initial contact and by communicating the initial project plan. An initial face-to-face meeting of all team members to discuss the project and to allow individuals to have an initial understanding of one another is one successful strategy to avoid conflict during the forming stage. Another alternative would be to consider the use of videoconferencing technology, should travel costs be an issue for the team. PC Webcams could be considered as an alternative to face-to-face or videoconferencing, the problem might arise regarding available computing power and bandwidth for all team members to allow this technology to operate effectively. PC Webcams might be an even more cost-effective solution; however, there are technological considerations for all team members regarding this potential solution (Hass, 2009). The project manager must link the introductions along with the communication of the project plan because pleasant group contact will not decrease intergroup tension. If conflict does occur, the project manager should act as a mediator to resolve conflict.

Group conflict in the storming phase is common (Tuckman, 1995). As individuals are learning their role and responsibilities, they will usually make mistakes. As communication is slower and less robust in the virtual environment, individuals may misinterpret these issues. Therefore, a project manager should plan on resolving conflict between team members. In the storming phase, a project manager must expect people to disagree, as individual responsibilities become clearer.

According to Project Management Institute™ (PMI) (2008), in order to mitigate this conflict, a project manager must take steps to clearly define expectations, expected tasks, deadlines, and ramifications. The better individuals understand their roles, the less likely conflict will occur. Additionally, when everyone is aware of what is required of each individual on the team, there is less of a concern that some people are performing below expectations. When tasks are agreed upon in the initial project meeting, there are fewer points of conflict due to ambiguity.

Another recommendation is for the project manager to use a project plan to manage the virtual aspect of the team, as well as creating a project plan for the team goal. In effect, the project manager has to manage two projects while leading a virtual team. By creating two separate documents, it helps outline the process and expectations of the virtual team, but it also creates a performance document where team members can anticipate potential problems in the virtual environment. This document will also help keep the goals of the team in focus while explaining the challenges that might arise. Team members can then become committed to the project as well as being accountable for their actions.

Practical Tip: Working across boundaries. A virtual project manager should ensure that the team respects the various worldwide time zones by scheduling conference calls at times that ensure no one has to endure strange hours more than others.

Practical Tip: High expectations. The virtual project manager holds team members accountable for schedules and keeping individuals informed.

Practical Tip: Training and technology. The project manger will ensure that team members have the appropriate training for all aspects of the project's technology. Prepare a slide show that is distributed to all team members regarding what training and tools are available. This is also a point to revisit several times during a project.

Practical Tip: Support of customers and stakeholders. The project manager will keep the customers and stakeholders informed of all risks and ensure that they provide the proper resources to the team members.

Virtual project managers must understand the team stages of a virtual project. Understanding this process allows a project manager to anticipate the period most likely for conflict. Negative conflict results in both wasted time and lost productivity for a project because conflict costs the organization money and is a contributing cause of project delays. It is estimated that the cost of conflict, when including ineffective managing of interpersonal situations, conflict avoidance, and lost project days accounted for $20,000 per employee, per year in addition to 20–25 percent of a manager's time was spent dealing with team disagreements (Johnson & Johnson, 2000). The virtual project manager must identify potential periods of conflict as well as understand strategies to cope with destructive conflict.

> Practical Tip: Handling destructive conflict. Addressing conflict immediately is the best method. Do not be afraid to schedule several sessions to repair the damage. Also, follow up afterward to make sure that things are operating smoothly.

The virtual project manager should ensure that all team members have compatible technology as well as, training on working with different cultures. The virtual project should quickly organize databases that allow sharing and learning. A project manager must review the available resources to make sure the appropriate technology is available to all members.

> Practical Tip: Technological compatibility. Make sure that everyone on the team has compatible technology. The project manager must ensure that each member of the project team, including clients and subcontractors, have compativel e-mail systems prior to starting the project. Send test e-mails or check team members; access to vital programs. Making sure that all technology is compatible in the beginning will save a lot of time later as well as help avoid strife and problems that can arise from having communities of haves and have-nots.

CONFRONT CONFLICT QUICKLY

Regardless of the type of conflict or how the conflict came about, the PM must confront conflict quickly. This will lead the team to cohesiveness and to approach the norming phase. As the team shifts into the storming phase, the PM should slowly shift the leadership to encourage team building and holding individuals accountable for deadlines and deliverables. The PM should be aware that he or she might have to remove disruptive individuals from the team (Curlee & Gordon, 2002; Lipnack & Stamps,

2000). During the performing stage, the PM should shift the leadership focus to removing barriers that the virtual team may encounter at the "home office" or with bureaucracy on the team.

The project manager should confront conflict quickly by scheduling a meeting to address it. If the team is in different time zones, it should be done at a time convenient for everyone. The team may dislike the disruption; it will send an important message that conflict is important enough to warrant a special meeting. Also, scheduling it at an off time will also show that the issue is critical enough that everyone needs to handle it in addition to the project. Successful handling of the conflict will lead to better team cohesiveness.

> Practical Tip: No one is comfortable with conflict, so address it as soon as you are aware of it. Putting it off only makes it worse and can make the conflict even larger. People have a tendency to make the negative larger than what it really is.

CREATE A CULTURE TO ADDRESS CONFLICT QUICKLY

"I feel the need, the need for speed."

—Maverick, *Top Gun*

Organizations rise and fall by how flexible and how quickly they can manage their processes and how well they can manage their customers. Organizations that are leveraging complexity should be more than able to address issues quickly. This applies to every aspect of the business; one of the most important aspects is the addressing of conflict. Conflict can sap an organization faster than almost anything else can. When people are not managing their time to manage the project, they are wasting time and in turn this waste becomes something that will destroy an organization.

As discussed prior, complexity is about trust. Trust is about people doing the right things at the right time. Trust is about utilizing resources appropriately and making a difference in an organization. So how does this apply to a culture? Consider the following:

- Organizational trust means that team members respect each other and communicate appropriately with each other.

- Individuals who trust one another will do their best to help one another.
- Teams that trust and respect one another will want to make processes as smooth as possible.

If the organizational culture has the above elements, then consider that the culture is supportive of rapid change and can handle addressing conflict quickly. The conflict might not even be major, but it is worth the time for the project manager to address the conflict quickly. Trust is made when people communicate and the lack of communication often breeds distrust. In fact, all the communication does not have to be positive to build trust. Conflict can detract from a project and cause systems to break down. Worse still, it can cause additional unforeseen problems as individuals attempt to work around different individuals (Montoya-Weiss, Massey & Song, 2001).

The culture should support the rapid addressing of conflict and the best way to handle this is to make it a priority and to make it an example. If conflict arises, the project manager must act swiftly to make the correction. The project manager must take the appropriate steps to correct the problem and to clear up the communication. Because most conflict is based upon miscommunication, it is more important than ever to open communication between individuals (Bass, 1990). Regardless of the action, the action must be swift—this is the example that one needs to set in order to be the model for future problems.

If an organization has a history of not addressing conflict, then one can expect that others will try to push conflict aside. Often managers will hide behind being too busy to address the problem or will assume that subordinates will take care of the problem. The fact that the information about the conflict has reached the project manager is a good indication that the problem will not go away on its own. Keep in mind that the faster the action taken by the project manager, the better, because the conflict has probably already been brewing for a while. In general, the following culture activities should be considered in order to address conflict swiftly.

- Get problems off the team.
- Banish conflict.
- Grow the team toward success.
- Manage the scope.

Get Problems off the Team

The more efficiently the project manager can deflect and address problems before they reach the team level, the better off the project will be. The goal of the project manager is to allow team members to handle their daily jobs. New problems may come down and force the team to do more work, often in the form of new reports, additional project details, and other potentially time-consuming activities. The more the project manager can handle those kinds of issues, the more the project manager will be creating a culture that addresses conflict quickly (Curlee & Gordon, 2002).

The more work that is pushed down, the more likely that conflict will arise. By avoiding adding new complex problems into the project team, the less likely one is to create new conflict. There are always times when the project manager must ask more of the team, but make sure to do it equitably. If one team member feels he or she is carrying more of the project, the more likely that conflict will arise.

Banish Conflict

If conflict arises, be sure not only to address it quickly, as already mentioned, but be prepared to banish a team member (Maznevski & Chudoba, 2000). It does not always come to this, but if a team member is a problem for the team, it is better to do without that person's skills than to drag down the entire team. One would be surprised at how much damage a single negative individual can do. The project manager must be mentally prepared to change the team if necessary. Too often project leaders wait too long for this option. Never wait—take action. The reality is that the leader must think of this as a substitution and not a failure. After all, if the skills turn out to be critical, the project manager can always call that team member back. More than likely, the team will come together more to avoid having the troublemaker come back, so one can expect that calling back a team member will actually be very rare.

Grow the Team toward Success

Make the team more successful by allowing them to showcase their skills. Nothing keeps conflict at bay more than success. If the team feels successful, any conflict that arises gets taken care of quickly—almost like magic. A good team knows that staying a top team is important in any organization. A team will often go to great lengths internally to maintain its reputation. This can help keep the team on the right track. Putting the right people in the right roles is critically important for the project manager as well as for the project (Rigby, 2009).

A team that has allocated its human resources in a manner that is effective will ultimately be efficient. Putting people in the right place and keeping them in the right place is what places teams on top. It might mean moving and changing people into different positions but it will keep things fresh and will make the project exciting (Godin, 2008).

Manage the Scope

Scope creep in projects is the bane of the project team. The more a project grows in deliverables, the more likely there will be conflict. If more is expected, one can be certain that there will be new variables in the project. When unaddressed variables exist in a project, the more confusion will develop in the process. Of course, if the project team has built flexible processes, then there are fewer problems, but keep in mind that conflict is about people, not processes. People can perceive this scope creep as more work for less money. This can drive even the strongest team to question their worth and to wonder if the team will share in the new value of the project.

The project manager needs not only to control the expansion of the scope, but also to make sure that they fight for increased compensation for the project. This does not always mean money, bonuses, or promotions, but it does mean that the team must gain something of value in exchange for doing more. After all, the team is not only doing more, but they are accepting more risk. The greater the risk, the harder the project, and in turn there should be some kind of reciprocal benefit to the team. A wise project manager must keep this in mind and advise the team of this newly earned value. If the project manager can keep bringing in more value, the team will do its best to mitigate conflict in order to avoid missing future benefits. After all, the benefits will cease if the project fails to meet deadlines or is perceived to be ineffective or inefficient.

Thus, the project manager must learn to create an atmosphere that can deter conflict, but also create a culture that it is addressed swiftly and appropriately. Breaking up a broken team is better than trying to win with a broken team. Hence, one must do everything possible to build a better team, even if it means changing the people when necessary. A successful project team member yesterday does not mean that the team member will be successful today. The project manager must learn that repeatability is important, and often times repeatability means making adjustments to a team to make it better rather than allowing one to be limited to what one had been in the past (Rigby, 2009).

CHAPTER SUMMARY

Virtual project managers must understand the team stages of a virtual project. Understanding this process allows a project manager to anticipate the period most likely for conflict. Negative conflict results in both wasted time and lost productivity for a project because conflict costs the organization money and is a contributing cause of project delays. It is estimated that the cost of conflict—when including ineffective managing of interpersonal situations, conflict avoidance, and lost project days—accounted for $20,000 per employee, per year, in addition to 20–25 percent of a manager's time spent dealing with team disagreements (Johnson & Johnson, 2000, p. 337). The virtual project manager must identify potential periods of conflict as well as understand strategies to cope with destructive conflict. The more that the project manager can anticipate and address, the smoother the transition will occur.

The virtual project manager must not only anticipate these conflicts, but they must communicate that they are part of the process. So when they arise, the project manager can address them swiftly in order to ensure they cause the smallest amount of disruption to the project. The more effectively the project manager can address these virtual conflicts, the more efficient the team will be. A veteran project manager understands that their value is made by what value they can add to the entire process. The more they can get the team to follow the plan and to work toward the stated objectives, the better morale will be and the greater the effectiveness of the team. All projects will not be successful; people will remember the well-run projects and will want to work on similarly led projects in the future.

CASE STUDY: CONFLICT AT PROJECT MANAGEMENT TECHNOLOGY, INC.

Jack Smith has been with Project Management Technology, Inc. (PMTI) for the last 18 years and has been involved with many different projects.

Jack has specialized in the big, expensive projects and he is often accused of wasting resources on a project and having his favorites on a team, but he always delivered big results to the client. He is also effective at meeting a tight deadline by aggressively forcing people to achieve often unrealistic deadlines. The largest complaint people have about him is that he is a ruthless taskmaster who has no concern for other people and is only focused on results.

In the most recent project, John met with the client on his own and negotiated a completion bonus for the project if it could be delivered 30 days ahead of schedule. When originally contracted, the project was for six months with fairly clear deliverables. Within the first three months the client has had over 150 change order requests and the primary focus of the project has changed at least three times.

Jack has already come to you to tell you about the great bonus that he feels that his team will be able to deliver. He is supremely confident and feels that the client is sincere about the offer and Jack has almost gone out and said that the bonus would be a sure thing and that he feels that a portion of the completion bonus should be distributed to him and his team. You asked about all the change orders and the lack of focus by the client, but Jack dismissed that, saying that the client's representative is new and is being driven by different forces within their organization. Jack feels that things have calmed down now with the client and points out that there have not been any change orders for ten days. In principle you were in agreement about his presentation of the facts, but shortly after your meeting with Jack, the whole project team contacted you with an entirely different picture of the project.

The project team consists of six other hard-working veteran employees and they have come to you, as Jack's boss and project executive champion, to talk to Jack about his most recent discussion with the client. Three of the team members are currently on site in Houston, Texas, and the remaining team members and Jack are located in the corporate office of Project Management Technology, Inc., in Jacksonville, Florida. Jack has spent a lot of time flying back and forth between Houston and Jacksonville and they feel that he does not see the day-to-day challenges that the team has to endure with this particular client.

The project team has come to you with certain grievances about being overworked and underappreciated because the client has been particularly difficult about accepting deliverables, even when they clearly meet the requirements. The client has been highly adversarial and has not treated team members with respect in meetings and only seems to tolerate Jack as the project manager. The team feels that they were not consulted in the recent negotiation and they would never have agreed to the completion of the project 30 days ahead of schedule; the project is barely on schedule now with three months remaining. The project team also feels that the client will never accept the project early even if they meet the required deadlines, because the details for acceptance have been too vague or are no longer valid due to all the changes. The team feels that the client is making a ploy to have the project done early for their own means and have no intention of paying out the bonus. They also fear that the client will present many last-minute change orders in order to force the project to be late and miss the window for the bonus.

It is clear from the meeting that there is a certain level of conflict between the project team and Jack as well as conflict with the client, and it must be addressed. Given that you do not want to create more conflict within this situation, consider the following questions:

- How would you address the conflict that has come to your attention between Jack and his team?
- How would you address the conflict between the project team and the client?
- How would you address the conflict between Jack and the client?
- Would you advise the project team of Jack's plan to distribute the completion bonus to the team?
- How would you address the problems faced by the team members on site and those who are remote?
- What coaching advice would you give to Jack in order to help him complete this project?
- Would you consider releasing either Jack or any team members from the project?

CASE STUDY REVIEW

The first question to consider is whether the actions taken by Jack are in accordance with the organization and the team's culture? If his actions are in accordance with accepted and practiced norms of behavior, then one must deal swiftly with this situation. There are two paths for this kind of situation.

The first path is that if his actions and behaviors are acceptable, then this should be treated as a case of a conflict intervention. The project manager (you) should attempt to conduct a face-to-face intervention. If that is not possible, then conduct it by phone. Never try to resolve a conflict like this via e-mail. Addressing something as delicate as this is difficult enough, so use the most robust communication available. Also, it is important to follow up after the resolution to make sure the situation has been resolved to everyone's satisfaction. A virtual team leader must keep in mind that individuals from different cultures might have a different perspective on conflict and conflict resolution.

Furthermore, the project manager must follow up with the team. Communicate to them as much as possible regarding the situation and how it is being addressed. Failing to close this loop will lead to negative perception of the situation.

The second path is when the person is acting in a manner that is not congruent with the organization. At this point, it is probably better to sever the relationship with the person. It does not mean firing them or laying them off, but it does mean that they need to get transferred somewhere else and fast. The team will not likely want the person back, and trying to force them back will only make the situation worse. It is possible that the damage can be repaired but it will probably take more effort and energy and will likely cause many other people to leave before it gets better. This is a difficult decision but ultimately it would be the right one.

Thus, conflict is difficult and forces a project manager to make some difficult decisions. The more the project manager is ready to take those steps, the better off the organization will be. Regardless of the path, either way will be uncomfortable and difficult for a while.

REFERENCES

Bass, B. (1990). *Bass & Stogdill's Handbook of leadership: Theory, research, & managerial applications*. (3rd ed.). New York: The Free Press.

Brown, H., Poole, M., & Rodgers, T. (2004). Interpersonal traits, complementarity, and trust in virtual collaboration. *Journal of Management Information Systems, 20*(4), 115. Retrieved July 28, 2004, from Proquest.

Cascio, W. (2000). Managing a virtual workplace. *Academy of Management Executive, 14*(3), 81. Retrieved May 10, 2004, from Infotrac.

Curlee, W. & Gordon, R. (2004). Leading through conflict in a virtual team. Proceedings of the North American Project Management Institute.

Duarte, D., & Snyder, N. (2006). *Mastering virtual teams: Strategies, tools, and techniques that succeed* (3rd ed.). San Francisco: Jossey-Bass.

Godin, S. (2008). *Tribes*. NY, NY: Penguin Group.

Hass, K. (2009). *Managing complex projects: A new model*. Vienna, VA: Management Concepts.

Johnson, D., & Johnson, F. (2000). *Joining together: Group theory and group skills, Needham Heights*. MA: Pearson Education Company.

Lipnack, J. & Stamps, J. (2000). *Virtual Teams: People working across boundaries with technology*. (2nd edition) New York: John Wiley & Sons.

Maznevski, M., & Chudoba, K. (2000). Bridging space over time: Global virtual team dynamics and effectiveness. *Organization Science: Journal of the Institute of Management Sciences, 11*(5), 473.

Montoya-Weiss, M., Massey, A., & Song, M. (2001). Getting it together: Temporal coordination and conflict management in global virtual teams. *Academy of Management Journal, 44*(6), 1251.

Project Management Institute (Ed.). (2008). *A Guide to the Project Management Body of Knowledge—Fourth Edition*. Newtown Square, PA: PMI.

Rigby, D. (2009). *Winning in turbulence*. Boston: Harvard Business Press.

Tuckman, B. W. (1995). Developmental sequence in small groups. In T. Wren (Ed.), The leader's companion: Insights on leadership through the ages, pp. 355–359. New York: Free Press.

Chapter 10

Risk Management through the Lens of Complexity

Figure 10.1 A bird taking flight is probably one of the more spectacular events to watch. With little effort, a bird can somehow accomplish self-propelled flight. When reviewed scientifically, birds are seen to perform a complex set of variables to achieve flight. Flight is a task that humans could only watch in awe until the past hundred years.

When most project managers think of risk, they think in the negative. Moreover, many project managers, especially those new to the practice, those not versed in risk, or those with weak people skills, try to ignore risk. Risk as defined by the Practice Standard for Project Risk Management (2009), published by the Project Management Institute^{TM} (PMI), is "an uncertain event or condition that, if it occurs, has a positive or negative effect on a project's objectives" (p. 111). Furthermore, an emergent risk is "a risk which arises later in a project and which could not have been identified earlier on" (p. 109).

New project managers and weak project manager may be tempted to write the risk management plan, identify the top ten risks, and not think about risk management again. The process of risk management tends to be ambiguous and depends on coordination with subject matter experts and others within the project. Successful senior project managers understand this. There are tools that may help to quantify the risks and opportunities, and even Monte Carlo projections or PERT analysis will help the project manager determine the end cost and schedule (Project Management Institute™ (PMI), 2008). However, these projections are not reliable if the input is not based on historical data, subject matter expertise, or some other quantifiable, dependable, and defendable data.

All the information presented on risk is an attempt to quantify an ambiguous entity. A risk is something over which a person has no control. It is similar to the observation that Lorenz made about the atmosphere. Although it appears to be in chaos there is some regularity, as is seen with the butterfly effect (Wheatley, 2000). Along those lines, there is also the reaction to risk, which can be observed when an anthill or s beehive is disturbed. This is when the risk occurs. Both will be discussed below.

THE BUTTERFLY EFFECT

Edward Lorenz, the famous mathematician, measured atmospheric anomalies with the expectation that the anomalies were totally random. Much to his surprise, this was not the case. He repeated the experiment several times to find that the atmospheric disturbances or anomalies were actually not happening in chaos but there was some order in the randomness. When plotted, the distribution graphic of the order looked like a butterfly. Hence the name "butterfly effect." (Wheatley, 2000)

This complexity subtheory was taken a step further, as an illustration that a butterfly flapping its wings in Brazil creates an atmospheric disturbance that may lead to a hurricane on the East Coast of the United States, or conversely, could prevent a hurricane on the East Coast. According to complexity theory, the same event in the same place may result in vastly different outcomes (Pievani & Varchetta, 2005).

Project teams and organizations are open systems since both are composed of human beings. According to complexity theory, both would be nonlinear. With nonlinearity, the project team and/or the organization may do the same thing several times with different outcomes. This relates to risk (Weaver, 2007a).

Many organizations that are in tune to risk and use lessons learned and historical baselines for developing risk management plans, risks, and risk contingencies, understand that what worked on one project may or may not work on the next project. In fact, the astute project manager knows that even in the same company this is true (Weaver, 2007a).

> Practical Tip: At the end of any project, make sure to have a "lessons learned" session. This should be informal in presentation, but formal in documenting the learning. If a mistake is worth learning from, then there is certainly something valuable that should be recorded.

Normally, how the risk manifests itself will depend on the interaction of humans. This is a complexity subtheory, Complex Responsiveness Process of Relating (Weaver, 2007a). In fact, the risk may not manifest itself due to human intervention. As discussed in Chapter 1, there are strange attractors in complexity theory. Simply said, nature has a way of repeating itself, as does history. Since human beings are an open system and open systems are a part of nature and human beings are a part of history, humans will repeat themselves. Therefore, risks will be repeated, but the outcomes may be vastly different and may end as an opportunity at times (Pievani & Varchetta, 2005).

While project risk management oversight should not be rejected for complexity ambiguity, the project manager should be using the social network of the project and the organization to understand the dynamics of the ever-changing risk environment. The project manager should be influencing stakeholders, managing expectations, and communicating the vision through the complex dynamic of the organization. Through this leadership opportunity and understanding of the cultural dynamic, the project manager can lead the project team through the ambiguity of risk (Weaver, 2007b).

BEEHIVE/ANT COLONY DISTURBANCE

How do open systems deal with ambiguity? In nature, it could be said that creatures just do or they disappear. Most living creatures have one goal: to perpetuate the species. This may be by protecting the queen, as with the beehive or ant colony, or with other species it may be to produce offspring. The human race is different. The human race has reached a higher level on Maslow's scale of needs. So by and large, most are not worried about perpetuating the human race. (Yes, there are exceptions in those places struck by natural disaster, human disaster, war, and strife.)

Discounting the exceptions, this may not be true when we break the larger human race down into smaller organizations. Think about a company that is downsizing or outsourcing. Immediately the organizations that are affected go into self-preservation mode. Depending on the culture of the company it may even become disruptive and result in workplace clashes or even worse. Remember that downsizing and/or outsourcing are projects. Each of the items discussed is a risk. Would the project manager even have noted them as risks? Maybe the project manager had, but most likely not, unless it had happened before in the organization.

Several of the items happen on the fringes of the project. For example, in an outsourcing or downsizing situation, it is unlikely for violence to occur in the workplace. If it does, there is nothing for the project manager to do. Others who are far more prepared for this situation must handle it. Now that it has happened, should the project manager be vigilant for future risks of this sort? Possibly. However, the risks of it occurring are so slight that preparation for this situation does not make sense. Nonlinearity suggests that humans will do the same things repeatedly with different outcomes. A value add for the project manager here would be to suggest a training session to understand and identify troubled individuals in the workplace. This would eliminate the need for risk management on the project.

The beehive and ant colony learn to react quickly to the unexpected. The ants and bees do not have complicated contingency plans in place. When the United States was attacked on September 11, 2001, there was no contingency plan for this event. Once it was realized what was happening, organizations had to react quickly. Air traffic controllers worldwide worked together; the passengers of United Flight 91 worked together; the U.S. president, his cabinet, and Congress made decisions quickly that day. If individuals had waited to analyze all the information before a decision was made, the outcomes would have been different.

Most recently, on Northwest flight 253, a Nigerian National attempted blow himself up on landing in Detroit. There is always a risk that U.S. international flights will be targets of terrorists. The thought was that the risk was so remote it was on the fringes of chaos. The systems were allowed to take care of security. The last system in place here was the open system of humans onboard the plane. In this nonlinear event, the consequences were minimal to the passengers.

What does this mean to the project manager? The project manager cannot account or plan for all risks. The future is unknown and the future is unpredictable. Through the project manager's leadership he should guide the team through traditional risk management and the ability to react swiftly and confidently to the unexpected (Weaver, 2007a; Weaver, 2007b; Pritchard, 2005).

> Practical Tip: A project manager must consider different contingencies. It is best to make a plan for the most likely contingencies. A good project manager should also have a "crisis" plan to cover emergency situations. The project manager should plan on having at least one "crisis" drill in order to make sure that everyone is familiar with the plan. If one does not have a drill, then one is never really sure if everyone knows what to do.

CHAPTER SUMMARY

Ambiguity is the key theme to risk. Since humans are nonlinear in their relationships, they will repeat the same thing with different results. Project managers will continue to have to deal with risks and opportunities that will continually have to be planned for throughout the lifecycle of the project. In addition, the project manager will have to provide the team with the tools and confidence to react to the unexpected (Weaver, 2007a; Pritchard, 2005).

When an open system, such as a project team or an anthill, can react to the unexpected quickly and revert to the norm or to a new norm, it is more likely to survive. The quicker the project manager can help the project integrate the unexpected into the project or eliminate the distraction, the better the team will be. A note of caution: the project manager should not become a part of the problem. There are times when project managers compound the problem by interjecting themselves into an unexpected issue. Remember that the project manager is the leader to provide guidance and help set the course back to normalcy (Hass, 2009).

The project manager has to balance risk management between the traditional process and preparing the team to expect the unexpected. Project managers must not believe that by thoroughly following a project risk management plan that the unexpected will not happen. It will be because, as Lorenz and all meteorologists, know the weather cannot be controlled!

REFERENCES

Hass, K. (2009). *Managing complex projects: A new model*. Vienna, VA: Management Concepts.

Pievani, T. & Varchetta, G. (2005). The strategies of uniqueness: Complexity, evolution, and creativity in the new management theories . . . or, in other words, what is the connection between an immune system network and a corporation. *World Futures*, 61. Milan, Italy: Routledge Taylor Francis Group.

Pritchard, C. (2005). *Risk management: Concept and guidance* (3rd ed.). Arlington, VA: ESI International.

Project Management Institute (Ed.). (2008). *A Guide to the Project Management Body of Knowledge—Fourth Edition*. Newtown Square, PA: PMI.

Project Management Institute (Ed.). (2009). *Practice standard project risk management*. Newtown Square, PA: Project Management Institute.

Weaver, P. (2007a). Risk management and complexity theory: The human dimension of risk. 2007 PMOZ Conference Proceedings.

Weaver, P. (2007b). A simple view of complexity in project management. 2007 PMOZ Conference Keynote address.

Wheatley, M. (2000). *Leadership and the new science: Discovering order in a chaotic world* (2nd ed.). San Francisco: Berrett-Koehler Publishers.

Part III

Case Studies of Applied Complexity

Part III will review two successful companies and one company type that are currently applying complexity theory to their organizations. These organizations have achieved success with organizational structures that are considerably less rule based and more driven by successful virtual teams operating with dynamic rules and a clear vision. These organizations clearly have structure and rules and are not the typical top-down institutions where everything is governed by hierarchy and order. Furthermore, these organizations are not limited by geography, and organizational compliance is often based upon team and individual rewards, which relate back to the business. These organizations will serve as actual examples of how complexity theory can be applied to successful organizations. In addition, at the end of each case there will be a review of the success of these organizations and how each entity applies complexity in a different manner to achieve success. Clearly, these examples are not designed to offer all the answers to the successful application of complexity, nor are these considered the only examples available. However, these organizations do offer a variety of applications to illustrate that complexity is far more pervasive in organizations than one realizes.

Chapter 11

SEMCO (Organizational Complexity)

Figure 11.1 This cactus is beautiful but thorny. If you look closely you will see that each branch is unique but symmetrically similar to the others. This offers a natural formation of complexity where a single plant can be beautiful while embodying different elements at the same time. A cactus is also an excellent metaphor for SEMCO, which is a rugged yet unique organization that somehow thrives in a difficult business environment.

COMPLEXITY AS AN ORGANIZATIONAL MANTRA

Ricardo Semler, the owner of SEMCO, originally did not want to become the owner of his dad's company. In fact, the company nearly cost him his life. At 25, he collapsed while touring a manufacturing plant in New York. The doctor examining Semler said it was the most severe case of stress he had seen in someone so young (Semler, 1993). This brush with death made him realize that there must be a better way to run a business. He realized there had to be more to business than just making money. Individuals spend most of their lives at work, and he wanted to make the environment more of a community, where individuals enjoyed coming to work and still contributed to the bottom line. Semler saw that in traditional organizations, management treated most workers like children, as if they were unable to make decisions, untrustworthy, and incapable of critical thinking, among other things. However, these same employees went home and ran a household, raised children, had bank accounts, did volunteer work, and may have had side businesses. Semler knew the paradigm had to shift.

He wanted SEMCO to be based on honesty, integrity, and self-direction. Lessem would have described Semler's movement as an "agent of transformation" (Bennis, Parikh, & Lessem, 1996). Further, Bennis, Parikh, and Lessem (1996) describes taking a vision and moving it to action as a mythical journey, and this is similar to Semler's eventual image for SEMCO. However, as Semler's SEMCO is further analyzed, the aspects of contingency theory are seen throughout the corporation and, therefore, through to the projects.

At this point, many may be thinking, sure but is the company making any money? And if it is making any money, is it growing and profitable? Well, consider these statistics: "from 1994 to 2003, Semco increased its annual revenue from $35 million to $212 million" (Gandossy, Tucker, & Verma, 2006, Chapter 15). Incredibly, the company continues to grow at a rate of 40 percent a year.

HOW SEMCO IS SUCCESSFUL WITH FEWER RULES

Semler decided to implement a democratic/participative and type of leadership style. According to Bass and Stodgill (1990, p. 421), there is mounting evidence that in the long run a democratic and participative style of leadership results in more productivity. Employees also tend to have a better sense of self-satisfaction. The anecdotal evidence presented in the book *Maverick* (Semler, 1993) provides support for the research presented by Bass and Stodgill (1990). There is mounting evidence in the business world that continues to support the participative type of management. Dee Hock (1990) created the VISA credit card empire based on the chaordic concept, which is a blend of participative, chaos (Beck & Cowan as cited by Bennis, Parikh, & Lessem, 1994, p. 11), and stewardship (Block, 1993) leadership. Today both companies continue to excel. Each company is employee focused. Minimal rules and regulations exist, but adherence is demanded unless there is a specific need to change the rules and/or regulations.

How the SEMCO Web Site Describes the Company

A company based on innovation, Semco does not follow the standards of other companies with a predefined hierarchy and excessive formality. At Semco, people work with substantial freedom, without formalities, and with a lot of respect. Everybody is treated equally, from high-ranking executives to the lowest-ranked employees. This means the work of each person is given its true importance and everybody is much happier at work (SEMCO, 2009)

WHY RULE-BASED ORGANIZATIONS MAY BE ON THE WAY OUT

Instead of making lots of rules, Semler lets his employees exercise their decision-making power. For instance, he let employees choose their most appropriate type of air travel (Semler, 1993, p. 4). However, there was a built-in enforcement mechanism because travel records were available to all within the company. Everyone will know if someone flew first class, so all watched what they did. Someone going to quit might take advantage of that, unless they had a real love and respect for the company.

Employee development and paying bonuses and profit sharing are tough issues with any company. Semco was no exception. Semler faced one dilemma when he pondered, "Do we help others find their niche or only hire those who have already shown they have found theirs (p. 26)? They chose to not "stockpile talent" and decided that 23 percent of a unit's profits were left to the employees to split among themselves however they decided (pp. 139–140).

Semler talked about "Crime and Punishment—SEMCO style." They decided that they would always press charges in a criminal offense. But for anything less than blatant criminality, they were loathe to get involved. Semco does not believe in "slapping wrists" (p. 162). Once, with a case of marijuana smoking at a company fiesta, they chose to deal with the problem in generalities and not to punish, but to work so that it never happened again (p. 163).

> Practical Tip: Consider both the culture and the law when dealing with problems. The law offers requirements, while culture can offer guidelines. Do not confuse the two because the law is not flexible while culture can change over time.

ANALYSIS OF SEMCO AND KEY COMPLEXITY LEARNING

Semler did not know how he was going to get to his end state, but he looked forward to the journey and the twists and mysteries it would provide. He was the protector of the vision as trials and tribulations provided him with substance for reflection and error correction. He also took risk and was able to penetrate the old school and create the new vision, which is still ongoing. His reward is to see SEMCO flourish and/or survive when other industries are failing. His workforce is committed to the new community of SEMCO.

The evolution of the SEMCO learning organization guided its strategic choices and beliefs regarding sustainable development. SEMCO has made the elements of the learning organization an ingrained part of the culture. SEMCO has learned, through trial and error, that organizational learning is a competitive advantage that can generate great benefits.

Sensory knowledge is the observance of something rather than the learning of a truth. Reflection is the step where an individual, usually the leader, ponders the combination of their vision of the future with their observation of the present. In this step, an individual contemplates past events in order to create a better vision of the future. Tacit is the individual understanding of the learning. "Tacit knowledge is highly personal and hard to formalize making it difficult to share with others. Subjective insights, intuition and hunches fall into this category of knowledge" (Nonaka & Takeuchi, 1995, p. 8). Explicit knowledge is taking individually experienced relevant knowledge and transforming it into knowledge that can be shared with others who did not directly experience the knowledge-producing event.

To better understand how this all relates to SEMCO, each step will be reviewed and then compared to the SEMCO organization with respect to the learning organization and sustainable development. Humans accumulate data through sensory input. Information is received from the five senses and is then processed by the mind. These senses are the only means of gathering information from the environment. All this information is received and processed by the brain and the central nervous system. Sensory knowledge is similar to the theory proposed by Bertrand Russell regarding "acquaintance" knowledge. "Acquaintance" knowledge occurs when one observes a thing rather than learns a truth from the knowledge (Russell, as cited in Moser & vander Nat, 1995, p. 214).

Ricardo Semler documents his gathering of sensory information from his initial time at SEMCO. Originally, the organization consisted of a tough-minded management organization. Slowly, this organization moved from traditional hierarchical thinking to a flattened democratic process. However, unlike most managers that simply add more bureaucracy or create more stringent controls, SEMCO opted to move toward a more democratic organization.

The learning process is what the body senses and then the mind interprets. This interpretation is known as reflection. Reflection is necessary to achieve a deeper understanding of the data collected. Reflection leads to making connections in a broader context. Reflection is the method by which humans process and assimilate information. Learning comes from integrating sensory experiences into thoughts that can be communicated.

Reflection is the review of experiences in order to make a meta-connection of experiences. To some individuals, reflective knowledge is more commonly known as life experience. An example of reflective knowledge can be seen in baseball. A seasoned baseball player instinctively raises a glove to catch a fly ball in outfield. This reaction is a learned experiential response. The input is a bat that has hit the ball, which is now flying through the air. To avoid the ball landing on the ground, one must catch the ball in a glove. This skill of ball catching is based upon experience and is not instinctual. This meta-connection comes upon reflection of sensory input. The rules of a game may have some roots in a historical educational context. These rules are an artificial construct. No one is born with instincts about the rules of baseball. After this reflection of all the data collected, the mind creates an interpretation of the experience and from this interpretation finalizes learning.

Much the same at SEMCO, no one is born with the instincts to create a successful organization. Ricardo Semler made many mistakes along the way to success. *Maverick* is the chronicle of Semler's voyage toward making the tacit knowledge of how to create an organization of business entrepreneurs.

Tacit knowledge is defined by three distinct elements: *knowledge of experience, simultaneous knowledge,* and *analog knowledge.* Knowledge of experience is sensory learning. Simultaneous knowledge is the timing of the learning process. Analog knowledge is practice and repetition of actions (Nonaka & Takeuchi, 1995, p. 61).

"Simultaneous knowledge" is sensory knowledge with the element of timing. The senses can only determine time from the perspective of "here and now." It is the mind that can differentiate time, not the senses. SEMCO became an organization that created a sense of timing. It took SEMCO four years to make the biscuit machinery successful, and by that time they had a two-year backlog of production for this highly successful machine (Semler, 1993, p. 132). Hence, simultaneous knowledge is an understanding of sensory experience with a perspective of time (Nonaka & Takeuchi, 1995, p. 61). For example, we must first learn about the existence of shapes before we can learn trigonometry. It would be impossible to experience shapes and trigonometry simultaneously.

"Analog knowledge" is the insight gained through practice. Skills can be better understood by repeating the steps to success. All professional athletes practice in order to hone their skills. "Practice is the best of all instructors" (Publilius Syrus, as cited in Eigen & Siegel, 1989, p. 497). At SEMCO, training does not exist in a formal training sense through a human resources department. Individuals are encouraged to develop a five-year development plan, and then to determine what training would be necessary to fulfill that plan. The training is then discussed and approved at the weekly business meetings of the organization (Semler, p. 335). This new approach to practice is critical to SEMCO's sustained development. When an organization purposefully creates an environment where people can grow without leaving, then it is planting for a future harvest.

The final level of understanding is explicit knowledge, which is the ultimate expression of understanding. Explicit knowledge is the process of taking individual knowledge and converting it to transferable group knowledge. "To convert tacit knowledge into explicit knowledge means finding a way to express the inexpressible" (Nonaka & Takeuchi, 1995, p. 31). Explicit knowledge is more than issuing memos or barking orders, "It's never enough to just tell people about some new insight, rather a person must experience it in a way that evokes its power and possibility. Instead of pouring knowledge into people's heads, you need to help them grind a new set of eyeglasses so they can see the world in a new way" (Brown, as cited in Drucker et al., 1998, p. 168).

> Practical Tip: Consider the different types of knowledge that are leveraged every day. Each of us might not be a master of each type of knowledge, but each of us can learn how and when to use each of the different types of knowledge.

SUCCESS ANALYSIS

Semler listened and realized early on that he had to allow the complexity theory of everyday life to take hold in his company. He also had to allow trust to be the mantra of the organization. For complexity theory to survive, the people involved in it must be trusted and must be allowed to resolve the issues within the complexity of the situation. SEMCO's business complexity, and yes, even Semler does not know what businesses have spun off, what are core, what are new endeavors, and what are being closed. The business is constantly in a state of flux or some would say complexity.

For some, this apparent lack of organization would be devastating but for others it is the only way. Can you imagine having your salary posted for all to see? Furthermore, the employees know that you established your own worth to the company. However, this is also a double-edged sword. Performance appraisals are posted as well.

> Practical Tip: Do not wait for a major event to reflect upon your life. There are opportunities for complexity in all aspects of life. The more you look, the more opportunity exists. Do not stop just because the first time is not immediately successful. Learning a new skill takes time and practice.

CHAPTER SUMMARY

Semler did not change his leadership style quickly. It took a life-altering event for something deep inside him to say that there must be something better in life. Now his company spins off other companies without major business plans; in fact, at any one point he does not know how many people are working for him and how many are working for new companies that have spun off. Is this complexity or sheer chaos?

By all accounts his company is making money even though Brazil's economy is not doing well. The latest research and interviews with Semco employees show these individuals are happy and enjoy the leadership. They believe they are being treated like worthwhile adults and are valued for their opinion; no matter whether they are the janitor or the CEO. Semler is not the CEO at all times. The CEO position is rotated. Board meetings are open to anyone who would like to attend. An employee (equivalent to a blue-collar worker) is always a member of the board. This is complexity being used to help the business succeed.

REFERENCES

Bass, B. (1990). *Bass & Stodgill's handbook of leadership: Theory, research & managerial applications* (3rd ed.). New York, NY: The Free Press.

Bennis, W., Parikh, J., & Lessem, R. (1996). *Beyond leadership: Balancing economics, ethics and ecology*. Cambridge, MA: Blackwell Publishers.

Block, P. (1993). *Stewardship: Choosing service over self-interest*. San Francisco: Berrett-Koehler.

Drucker, P. F., Nonaka, I., Garvin, D., Argyris, C., Leonard, D., Straus, S., Kleiner, A., Roth, G., Brown, J., Quinn, J. Anderson, P., & Finkelstein, S. (1998). *Harvard Business Review on Knowledge Management*. Boston, MA: Harvard Business Review.

Eigen, L. D. & Siegel, J. P. (1989). *The Manager's Book of Quotations*. New York, NY: AMACOM.

Gandossy, R., Tucker, E., & Verma, N. (Eds) (2006). *Workforce wake-up call: Your workforce is changing, are you?* New York: John Wiley & Sons.

Hock, D. (1990). *The birth of the chaordic organization*. San Francisco, CA: Berrett-Koehler.

Moser, P. K. & vander Nat, A. (1995). *Human Knowledge: Classical and contemporary approaches* (2nd ed.). New York, NY: Oxford University.

Nonaka, I. & Takeuchi, H. (1995). *The Knowledge-Creating Company*. New York, NY: Oxford University Press.

Project Management Institute (Ed.). (2008). *A Guide to the Project Management Body of Knowledge—Fourth Edition*. Newtown Square, PA: PMI.

SEMCO. (2009). www.semco.com

Semler, R. (1993). *Maverick: The success story behind the world's most unusual workplace*. New York: Time Warner Books.

Chapter 12

Web-based Universities (Multilevel Complexity)

Figure 12.1 This stone building is a level structure made up of very similar yet unique stones. This spire represents higher learning found in complexity-based organizations.

Some would say that Web-based universities are a passing fad. Others say they are diploma mills. Still others specifically state how hard the classes are. Most would agree that online universities are here to stay and each online university offers its own unique style of learning. One thing they all have in common is technology. Some have been more creative than others. Whether you agree with the above premises of online or Web-based universities or not, you will agree that even the most traditional university now offer online or Web-based training.

The analysis offered here is not representative of any one Web-based university. Rather it is based on how several online universities have incorporated the principles of complexity theory. By doing this, these universities have achieved and maintained an accredited and thriving institution, whether it is for profit or not for profit.

COMPLEXITY AS APPLIED TO A WEB-BASED UNIVERSITY

Web-based universities must rely on technology for most aspects. Some of these Web-based universities may have a brick-and-mortar front for the administration, the administrative services, the IT team, and other services. Others will have a brick-and-mortar front for appearance and to accommodate a very small staff. A Web-based university will always have a need to have a server "farm" and help staff.

These universities also have faculty. Initially, the Web-based university normally starts with a cadre of professors. Prior to accreditation, it may be hard to recruit faculty and students. This is where the complexity theory starts to come into play, more specifically through the concept of six degrees of separation. The concept of six degrees of separation is that any given individual can be connected to any other, unrelated individual in another city, by no more than six steps. Hence, if one were to reach out to friends of friends that one could connect to anyone else on the planet. On average, this process would take six steps and hence six degrees of separation (Watts, 1999).

Faculty members will start to reach out to their networks, pre- and post-accreditation. These colleagues will continue to reach out to their networks. Eventually, the university will be able to staff with faculty. Students may be more difficulty to recruit prior to accreditation. Recruiting of students may be done through some hybrid of six degrees of separation; the university has to be somewhat flexible prior to accreditation.

Students are at risk of achieving a hard-earned degree that is worthless if the accreditation is not achieved.

The full-time faculty will be the foundation of the Web-based university and will rely on its corporate knowledge to develop the foundation classes. As new faculty is added to the university's staff, new classes will be added. The university's staff may have some outlines and suggestions for the course structure to conform to the technology or there may be none. Most likely there may be some to ensure the university will meet accreditation standards. On the other hand, it is not unusual for the staff to advise the new faculty member of the technology requirements and to write the class to meet those needs.

This may create various levels of rigor in the same class, a condition seen even in an accredited or a traditional brick-and-mortar university. Then how does an online university ensure accreditation? There is a tradition of academic freedom in most universities. There is a fine line between allowing faculty members to teach the academic subject as they see fit while still upholding the standards.

Universities then periodically "audit" the classes, review them if there is a complaint from a student, review them if the faculty member does not meet certain technology triggers, or the university may have certain scheduled peer reviews. The peer reviews may have very defined metrics or may be less formal. Each of these conforms to complexity theory in the sense that the faculty member does not know when they will happen or who is doing them.

A university relies on diversity. Each university professor brings a different perspective to the subject. The university must meet the strict elements of accreditation. There is the element of trust that the university affords its instructors; at the same time each university must understand how it is going to monitor that trust. The instructors must decide if that type of trust is acceptable.

HOW A WEB-BASED UNIVERSITY CAN THRIVE APPLYING COMPLEXITY

Web-based universities' leadership may think that it is monitoring the classes, but do they? Maybe their staffs are running reports to see if the professors meet the technology requirements; however, this does not mean the professors are really teaching. In reality, online university leadership is relying on complexity theory.

The leadership makes the assumption that the professors are doing their job; however, if there is a disturbance then there is action to be taken. The actions are similar to a project. A professor does not appear in a class; a professor does not meet the requirements on repeated occasions in a class; or a student disrupts a class. The entire university does not act, only the part of the university that deals with those professors who teach those classes deal with the issues.

Another aspect that must be looked at is the technology. A very popular online university had its entire server system collapse and so did its backup. This was catastrophic for the students, for the professors and for the university.

The class worked around the situation. It took the university almost a week to react to the situation, yet the students and professor determined a workable solution in only about 36 hours. According to complexity theory, it is easier to allow those involved to resolve the situation than to expect a bureaucratic organization to develop a solution.

WHY COMPLEXITY IS SO IMPORTANT IN A VIRTUAL UNIVERSITY

Complexity is central to most aspects of teaching and curriculum facets of the Web-based university. Without complexity, the Web-based university would have a difficult time of oversight. New professors are brought in at various times and are trained at different times by different professors.

These professors are normally mentored for their first class. Once these instructors pass their first class with a minimum qualifying score, the new professors either sign a yearly contract or a contract for each class. The minimum contractual requirements are outlined and the professor must uphold these standards. Without these standards, trust is broken. When trust is broken, complexity may then take over. It should be noted that this is not always the case. Also, it should be noted that a professor may slide with one or two classes and not meet requirements. The professor is on the edge of chaos or complexity but may not totally enter into the world of complexity.

As professors comply with the rules of the contract, complexity stays in check. When the professors start deviating from the contractual requirements or cannot deal with difficult students, the complexity starts to become significant and can be useful to the university. The professor may be involved with the staff from the university with contractual issues or the staff may have to determine the issue without the professor. For example, a complexity resolution might be to replace the professor. While this is disruptive to the class, the new professor who is used to the complex online environment will adapt to the complexity of the environment.

Technology is fundamental to the online university. We also know that fundamentally technology will eventually fail. Whether the university gives the professors flexibility or not, professors have to be willing to be flexible in the world of technology. There are students in areas of conflict, students who are facing natural disasters, the university itself may have a catastrophic event, or the professor may have a technology issue. Each complexity issue will define how it is addressed. Complexity based-solutions often evolve to meet the needs of the organization just as an organization with a broken leg will learn to cope. A complexity-based system will create a solution that can work.

> Practical Tip: Technology is very important to communication and project management today. Understand the technology that is in use by everyone on the project team and make sure that is always a backup plan to replace technology temporarily. It is not a matter of *if* technology will fail; it is a matter of *when* technology will fail.

ANALYSIS OF A WEB-BASED UNIVERSITY AND KEY COMPLEXITY LEARNING

Web-based universities have learned to embrace complexity as a matter of survival. Initially, online universities struggled to keep faculty and staff members in a traditional brick-and-mortar environment while students were in the online environment. The universities found this to be cumbersome and extremely expensive without adding to the quality of the program. Slowly university staff embraced keeping at least the adjunct faculty in a remote environment.

Over the years, university leadership came to realize that traditional policy and procedures do not work with remote faculty (tenured, full time, or adjunct) and students. Through trial and error, most Web-based universities realized that something different had to take place. As a result, these universities give more leeway to the faculty, allow leeway for technology, provide structure as needed, and provide enough oversight to ensure quality but not to stifle academic freedom and creativity.

The administration in many of these universities has allowed their faculty much liberty. Yes, the faculty signs a contract (either yearly or with each class) and must abide by a faculty handbook. This handbook normally outlines the minimal classroom policies, academic rigor, and faculty conduct required to maintain employment at the university. Also, some of the faculty that maintain best practices are invited to mentor or teach new faculty members, do peer evaluations of existing faculty members, and write or update courses. After training, the faculty member is basically on his or her own. Many universities may provide either a program chair or a team of individuals to reach out to if there are issues in the classroom but by and large the faculty member is left on the rim of complexity.

As mentioned previously, a major online university had a major technology outage. Most online universities will encounter this at one time or another and the technology outage generally lasts an hour to several hours. What made this technology outage different was that it lasted over a week. Neither students nor faculty could communicate with one another (except via personal e-mails) or with the administration. This created the ideal situation for a complexity intervention. The university provided direction after eight days. For many classes, this was unsatisfactory because the class was already complete.

Complexity theory states that although things appear chaotic and random at the periphery of the system, there are patterns. Since most of the faculty members had alternate e-mails for the students, the faculty had worked alternate arrangements with the students. Sometimes it was within the university guidelines and, presumably, sometimes it was not. In this case, ethics and integrity and the trust established by the student and the faculty member and due to the unusual circumstance trumped the university guideline. The operative word here is "guideline." (Cooke-Davies, Cicmil, Crawford, & Richardson, 2007).

Technology failure resulted in a complexity scenario above and drove the faculty and university staff in various directions. Most likely none were wrong, but depended on the lens and the information available. Each university has a diverse manner in handling technology. The university will establish its technology based on the mission of the university and how technology intensive the classes are. Some universities present classes that are broadband intensive while others are more focused on online texts and papers with no group classes or broadband-intensive environments. While each environment could present complexity opportunities, those with actual group classes most likely would have the most opportunity for complexity-related to technology.

Academia, especially for tenured professors, has always prided itself on giving professors academic freedom. Most online universities pride themselves on providing their students with faculty that have current professional business credentials. This presents an interesting dilemma for the university. A business professional is not an academic but does have a lot to offer a student, yet it must be done with an academic overtone. As such, this new style of professor must be provided just enough structure to be a professor. The university provides various types of training, and the structure varies depending on the training. Remember the curriculum structured for new-professor training is developed by online professors but the instructor teaching it most likely has a lot of leeway.

Other policy and procedures to help the professor may include a handbook, a blog, a contract, updates throughout the year from the provost/dean, training or college meetings to provide policy updates, and other means of providing news to the faculty. Some of these may be required, while others are highly recommended or suggested. This is done via email communication or via a portal. Faculty knows or should know that they are required to meet the requirements of the school. If not, normally the faculty member is reminded by e-mail that he or she is not complying.

Those who do not comply, either in the classroom (which are monitored by technology and by student complaints) or through other requirements, are followed by university staff. In addition, university leadership will also ensure that professors follow minimal requirements. Each university has different methods but most have a peer review and some software methods. In other words, someone from the faculty or staff will review a class once a year.

SUCCESS ANALYSIS

Web-based universities, whether for profit or nonprofit, must have enough students enrolled to be sustainable. In addition, the university has to be able to provide a diploma that is meaningful to the student. This means that the diploma is backed by an accredited university. Web-based universities have a large hurdle to cross when starting, especially those that are starting without a traditional college or university behind them.

The Web-based university that has the traditional university behind them normally will not have to worry about accreditation. However, they still have a hurdle to face. Most universities' leadership and tenured professors are steadfast in their approach to how academics should be approached and how tenured and nontenured professors should be treated. Complexity theory would be foreign to the newer Web-based universities.

The newer Web-based universities, and those from traditional universities who understand the need for flexibility, understand the need for academic rigor while still allowing academic freedom in the classroom. The successful Web-based universities have embraced complexity theory. This was done through the need for survival and through trial and error.

A successful Web-based university that has seen double-digit growth originally was authoritarian in its classroom oversight, in adhering to the class syllabi, and in the length of feedback given to students on each assignment. This university's administration soon found that the overhead to implement this approach was too expensive and in the long run did not make sense. It slowly adopted a hybrid complexity approach.

This hybrid approach is similar to what a seasoned project manager would do on a project. Initially, a new faculty member is trained and mentored through his or her first class. Then once a year the faculty member is peer reviewed. The faculty member will no longer hear from the university regarding teaching classes unless established metrics are not met or there is a complaint that can be verified (Singh & Singh, 2002). Only when something happens on the fringes of chaos may a member of the university staff need to become involved. Even then the faculty member may be able to resolve the issue himself or herself.

The faculty member also has to be able to be able to contact appropriate parts of the university when there are different issues. Each Web-based university handles this aspect in a variety of ways. Some have program leaders logged into various types of instant messengers. These leaders are available to answer questions at a moment's notice. Other Web-based universities handle the issues via e-mail.

CHAPTER SUMMARY

Web-based universities will inherently function better when certain aspects are allowed to function on the fringes of complexity. Even during the start-up process these universities rely on certain aspects of complexity, such as the social networking to find faculty members. The staff at the university may also depend on the same phenomenon to spread the word about the university.

The traditional rules-based environment does not work well in a virtual-based setting. The rules would be hard to enforce and would require a large overhead of staff that would be cost prohibitive. Therefore, allowing technology to monitor certain metrics and doing the minimum to maintain certification is where most Web-based universities tend to keep the rules. Then the rest of the issues are handled as outliers or those on the fringes of chaos (Jaafari, 2003).

REFERENCES

Cooke-Davies, T., Cicmil, S., Crawford, L., & Richardson, K. (2007). We're not in Kansas anymore, Toto: Mapping the strange landscape of complexity theory, and its relationship to project management. *Project Management Journal*, *38*(2), 50–61.

Jaafari, A. (2003). Project Management in the age of complexity and change. *Project Management Journal*, *34*(4), 47–57.

Singh, H., & Singh, A. (2002). *Principles of complexity and chaos theory in project execution: A new approach to management.*

Watts, D. (1999). *Small worlds: The dynamics of networks between order and randomness.* Princeton, N.J.: Princeton University Press.

Chapter 13

Small Team Complexity

Figure 13.1 Examine all the elements of this view of the beach: the stones, the clouds, the ocean, the plants, and the lone tree grouping. Each element exists alone, yet they all interact in the picture. Consider each element as its own small team but together they create a larger vision.

WHOLE FOODS™: A DIFFERENT KIND OF MARKET

To understand how Whole Foods™ is a different kind of market, there must first be a short introduction to the company's early years and growth. These early years ingrained certain ideas into the culture of the organization. Although some organizations that survive early difficulty manage long-term success, some interesting problems tested the mettle of those involved in Whole Foods™.

The Whole Foods Market™ story began in Austin, Texas. The company was founded by four local business people who decided that the natural foods industry was about to mature to a point where an entire supermarket could be dedicated to natural foods. Other businesses had natural foods offerings, but there were no more than six natural food markets in 1980. Whole Foods™ began as a merger of Safer Way Natural Foods and Clarksville Natural Grocery. The first Whole Foods Market™ opened in Austin in 1980 with a staff of only 19 people. It was an uncertain future; however, the business turned out to be very successful from the start. All seemed like a certain path to success until Memorial Day 1981. A terrible flood hit the city of Austin on that day and did almost a half million dollars worth of damage to the store and inventory of Whole Foods Market™. At the time, Whole Foods Market™ had no insurance to cover the damage. Most start-up companies at this point would close their doors and give up. However, Whole Foods Market™ is a different kind of market.

Since its inception, Whole Foods Market™ was distinctive in two ways that has separated it from the competition. First, Whole Foods™ has remained firm to its values and principles. Second, Whole Foods™ has always been about community and teams. These two elements are precisely why Whole Foods™ was able to survive this initial challenge when the worst flood in 70 years destroyed much of Austin. Once the floodwaters receded, customers and neighbors voluntarily assisted the staff to restore Whole Foods™. Furthermore, banks, suppliers, and investors were patient during the restoration and within 28 days after the flood, Whole Foods™ reopened.

This flood did more than help the company survive a difficult time; it defined the organization moving forward. The founders understood that a community will make or break a business. There is a definite difference between being a servant of the community and being part of the community. Being a servant of the community means that if the business stops offering what the community desires, the community will move to the next business. Being part of the community means that if a community member is in need, then community will come to their aid. Understanding this difference has crystallized the ideas of values and principles as well as community and teams into the Whole Foods™ way.

Values and Principles

Values, particularly corporate ones, are not new and continue to be a source of discussion among business people in all industries; there is a difference between what happens at Whole Foods™ and what happens in many other corporate cultures. Current corporate theory contends that when many people within a business share values, the organization will then achieve some form of nirvana and profits will follow. This model is too simplistic and there must be farther-reaching implications for the values than just having people share them.

Organizational theory identifies values that can both empower and shackle an organization. Values that are universally known within a company are values that will be followed at all levels of the organization. Without these values permeating the organization, these values will be ignored at some point. "In fact, six out of ten large U.S. firms today have developed a specific statement of values, otherwise referred to as a philosophy, credo, vision, aspiration, mission or set of principles" (Ledford, Wenderhof, & Strahley, Winter 1995, p. 1). Yet the same authors state that, in their experience, "the typical corporate philosophy statement has, at best, a negligible impact on attitudes, beliefs, and behavior of organization members" (Ledford et al., p. 2). The most obvious problem with corporate values is that there is no connection between organization reward and corporate values. Only when organizational reward is directly linked to values will people start religiously following the corporate values.

This is part of the difference at Whole Foods™. Organizational values are not only linked to the reward system; organizational values are the reward system. Whole Foods™ adopted core values since their inception and have remained true to these values.

All Whole Foods™ associates believe in and live by these ideals (Whole Foods™ Web site, 2009):

- Sell the highest-quality natural and organic products available.
- Satisfy and delight our customers.
- Support team member happiness and excellence.
- Create wealth through profits & growth.
- Care about our communities and our environment.
- Create ongoing win-win partnerships with our suppliers.

There has been considerable research into the creation of a strong positive culture based upon a shared value system. When employees embrace the corporate values, these become what are known as First Order Corporate Values. First Order Corporate Values, such as those found at Whole Foods™, increase an organization's likelihood of surviving and achieving business excellence (Giblin & Amuso, 1997, p. 4).

Employees also need to know that the managers and leaders respect the personal values of the employees. In essence, corporate and personal values "in the true sense, are basic, fundamental, enduring, absolute, and irrevocable" and form the fibers of its existences (Giblin & Amuso, 1997, p. 1). When properly executed within an organization, corporate values give stakeholders a sense of belonging to something larger than life.

An integral element of embedding these values into an organization is to create a direct link between the values and the overall performance of the organization. Each employee has a stake in the overall performance and financial value (Collins & Porras, 1997, pp. 54–70; Dahlgaard & Dahlgaard, 1998, p. 5). When the link is there, it offers the opportunity for every individual to make a difference. Furthermore, world-class companies like Motorola™ and 3M™ are able to use their corporate values effectively for the benefit of the stockholders and stakeholders.

Principles, like values, are important to the organization. Corporate principles are more of the guiding principles of organizations. These are often matters that affect an individual but are much broader. Within these organizational values, there are two principles that are not directly expressed; however, both of these principles are important to Whole Foods™. These two principles are trust and equity.

Trust

Trust at Whole Foods™ is a reciprocal relationship. On one side, the employees know that Whole Foods™ will do its best to make sure that they are taken care of. On the other side, Whole Foods™ knows its employees will do all they can to make the company successful. Whole Foods™ associates understand that trust is something that is shared at all levels. The company is successful if the associate is successful (Stringer, 2009). If the associate has meaningful engagements with customers, then the customers will return and will make the organization successful.

In the business of groceries, there is no such thing as a onetime purchaser who will not have any significant impact on the organization's success. Success comes from building a relationship of trust with a client who understands the value that Whole Foods™ brings to the grocery shopping experience (Kasdon, 2007). Individuals will purchase groceries continuously for decades so the goal of Whole Foods™ is to become the grocery store of choice for all who come through their doors. Earning that loyal customer is about trust. Trust the prices will be fair, trust the quality will be high, and trust the service will be available and accurate when needed. Whole Foods^{TM} associates understand that a delicate balance exists and that relationship is something that needs to be continually reinforced. There is so much competition that just because Whole Foods^{TM} did a good job last week does not guarantee that customers will return this week. Greatness is not about a one-hit wonder; it is about a continuous relationship of trust.

> Practical Tip: Recognize and understand that customers are testing and reviewing trust on a daily basis. Consider how many times you have to feel cheated by a store before you stop going. It usually takes one bad experience to erase years of good experiences. If you found that the pricing at your local grocery store was vastly different from another proximate grocery store, you would consider shopping elsewhere on a regular basis.

Equity

Equity is an interesting factor at Whole Foods^{TM}. Many organizations can be successful for long periods of time without this element. Equity is about offering a correlation between contribution and reward. At Whole Foods^{TM}, there is a correlation between a team member's compensation and his or her performance. And this is not dictated by the organization, but by the team. The team (tribe) must agree to hire and agree to fire (if necessary) an individual as the performance of the small team determines the compensation for the team. So if a team performs, they are compensated for their extra performance. If a team does not perform, then the entire team does not achieve certain bonus payment milestones. Hence, there is a huge equity in compensation for a small group (tribe). This kind of compensation is rare, but when it exists it creates a feeling of equity as well as giving strong motivation to every individual to contribute as much as possible. There is no holding back in the Whole Foods^{TM} organization because the best thing a person can do is to contribute to the team.

All these elements contribute to making Whole Foods^{TM} a different kind of organization. There is good reason to respect the small team importance since an individual's compensation is based upon his or her contribution to the bottom line. The Whole Foods^{TM} example offers insight into what organizations will look like in the future and what can be done with scarce resources in a highly competitive market.

> Practical Tip: Consider doing something different in your industry. Too many times organizations are limited by what the competition is offering. Innovation is offering what others want and offering what others are not offering while remaining competitive.

The elements discussed so far do not always lead to complexity. Some highly successful organizations have not heard of complexity and many of those organizations continue to thrive. What is interesting about complexity is that organizations do not need to know they are applying these ideas in order to make them successful. Many times, organizations are applying complexity and not ever knowing it. The learning from this is that the most successful form of complexity comes not from a fancy label, but from making it part of the culture. It should not stand apart as an element or mission statement but it should be a part of the organizational expectation. Just as any good-quality company should never allow a defective product to leave its factory, a complexity-driven company could not conceive of the organization working in any other way. This is what makes success; as the saying goes, greatness comes from within. Complexity comes from within an organization.

HOW SMALL TEAMS APPLYING COMPLEXITY CAN MAKE A DIFFERENCE

Whole Foods™ has been applying complexity to its organization since the beginning, but this knowledge is now spreading throughout the world. In particular, small teams have already been seen to be both a boon and a threat. Small, determined teams able to move from design to an effective plan can certainly change the world. In some negative cases, it has already been shown that a small group of determined and misguided individuals can threaten large groups of innocents. However, small teams are certainly a force that can also bring order throughout the world.

Interestingly, as complexity is being resisted by societies of project managers, the U.S. military is realizing the importance of complexity-based strategies and the importance of small teams. In particular, there is a movement within the branches of the Army and Marines to deploy complexity in small team operations in order to allow these teams to be more flexible in meeting their goals. In particular, there is now circulated documentation within these branches of the military that codify how small teams can make a difference in counterinsurgency operations (US Army & Marine Corp, 2007).

The purpose of this new military doctrine is to address how counterinsurgency operations are to be handled. This new military thinking offers an insight and learning that has been completed over the past decade. Instead of creating another operational manual designed to describe different situations and the rules of engagement, the goal has been to create a dynamic manual that guides the operation toward success without offering specific plans for each and every situation. There is strong evidence that creating a path with only one answer is not the best use of the human resources available. Not every team or organization will have the exact same skill set or the same type of individuals, so trying to create a one-size-fits-all solution is rather limited(US Army & Marine Corp, 2007). It also makes for a situation where not all individuals will be able to achieve the same goals with the same results if the groups are not trained, mentally aware, or physically capable as all others. Human systems and humans in general have too much physical and mental difference in order to always offer the one successful answer to a problem.

Even more importantly, in times of military conflict, if there is only one correct response to a situation, the enemy will quickly be able to determine this response and react accordingly. For example, during World War II, the U.S. military forces in the South Pacific started with a strategy of island hopping in order to defeat the Japanese. The strategy was simple; U.S. forces would start with the islands closest to Hawaii and defeat the garrison on every island going toward Japan. It was thought at the time that a complete conquest of all island garrisons was necessary for victory. However, the Japanese quickly recognized this strategy, so they would fight tenaciously for every island, making the process particularly difficult for the U.S. military. Furthermore, once the garrison felt they could not hold out any more, they would sneak away and go to the next island to engage the U.S. military again. This made for very slow progress in defeating the Japanese; however, it became apparent over time that this strategy was not the most efficient.

The U.S. military realized the Japanese understood their strategy and so the military operation in the area was significantly hampered because the enemy was able to anticipate their next move. This causes a dangerous problem when encountering an enemy. If an enemy already knows opponent's next move, then they are able to prepare for it. Fortunately, the U.S. military forces at the time realized this problem and were able to modify their strategy to move away from the strategy of conquering every island to one that focused upon capturing strategic bases that were important for air- and sea-based attacks upon Japan. The U.S. military moved away from their original strategy of total conquest toward one of strategic conquest where only the necessary islands were captured and efforts were made to strand the remaining garrisons on the smaller, less significant islands. During this change of strategy, the Japanese were unable to adapt quickly in order to deflect the U.S. military's next move. The problem was that the Japanese wanted to hold on to as much territory as possible, and they lacked the resources to hold all their positions while defending against concentrated attacks. The new strategy put the Japanese at a strategic disadvantage as they were forced into a position to hold all the islands and to defend against specific concentrated attacks by the U.S. military. Without having an understanding of where the next attack would be, they could not marshal their forces as successfully as in the past. This new strategy was more successful, although it was still very difficult to defeat an entrenched and determined foe.

This was certainly an eye-opening experience for the U.S. military at the top level. Deploying a changing strategy that could take advantage of the tactical situation was clearly new for the military at that time. It is interesting to note that moving forward almost 70 years, the same thinking is being applied to small groups. Having small, flexible groups that can take advantage of the local situation is now understood to be the best manner of handling a situation.

The military needs to train a large number of individuals with different backgrounds and experience in a manner that allows all to operate together in a single focused purpose. However, the military organization understands the necessity of small teams that not only can operate in hostile situation but can be flexible enough to leverage the local conditions (US Army & Marine Corp, 2007). No longer is there one manner to handle a hostile opponent, particularly one that is entrenched and not favorable toward the occupying force.

Small Team Decision Making with Complexity

The new military view of handling small groups that must deal with counterinsurgency situations is to move in a continuum from design to

Design → Planning → Commander's Visualization → Intuitive

Figure 13.2 Small team decision making process

planning and from planning to the commander's visualization and intuitive decision making (Figure 13.2).

Design: A small group leader will be able to create a simple design in order to deal with conflict.
Planning: A small group leader will be able to plan out a simple strategy that leverages existing resources in order to achieve their given orders.
Commander Visualization: This is essentially the vision of the command that is grounded in the orders and values of the team.
Intuitive: This is the leader's intuition at work to update the plan based upon the situation on the front line.

> Practical Tip: Consider using the Design > Planning > Commander's Visualization > Intuitive model in a project. See if this small team decision making process can help improve the project.

This model allows for leaders to react quickly but also to utilize subliminal training, intuition, and expertise. Even the military recognizes that the best thermometer for a situation is experience and firsthand knowledge. People will instinctively know the right thing to do, but often are restricted from making that decision. By trying to remove those impediments, it can allow an individual to make good decisions based upon complex inputs.

Instead of having a step-by-step process to handle a conflict situation, the military understands the necessity in having flexibility in any plan or leadership decision in order to cover all the situations that may occur in the field. Instead of trying to suppress individuality, creativity, or leaders gifted with a particular talent, the military is offering a flexible rule book to assist in the struggle against hostile forces (US Army & Marine Corp, 2007).

Flexibility is now taking the place of rote process and procedure. Rather than having a set doctrine developed to address a particular problem or conflict, there is now a growing movement to create flexible rules that can be interpreted and implemented based upon the initiative of the leader on the front line. Instead of requiring a rule and seeking authorization, a frontline leader can expect to have a certain level of discretionary ability to cope with unforeseen situations that could arise. Individuals must learn to manage themselves in most circumstances (Drucker, 2001, US Army & Marine Corp. 2007). Rather than utilizing a high-risk solution, which could also be already understood by the enemy, new military leaders are given a degree of autonomy in order to achieve their given directives.

This will lead to high initiative units that will be able to decisively contend with new enemies and their updated tactics. Instead of being tied to orders or commands from a command structure that has never seen the battlefield, the military of the future will be able not only to address an abstract enemy but also be successful without being tied to a bureaucratic command system.

HOW COMPLEXITY ALLOWS ENTREPRENEURIAL TEAMS TO BE MORE SUCCESSFUL THAN RULE-BASED TEAMS

In the past, rules-based organizations have been the method to maintain order and structure to all companies. If something went wrong, then there were not adequate rules or sufficient structure to avoid errors. Organizations strive to eliminate errors and mistakes to a point where they are nonexistent. Machines have assisted in this quest to eliminate errors, yet humans continue to generate errors. In fact, humans have often been seen has the most error-causing aspect of any business. The problem with this concept is that there is only one 100 percent proven method to avoid errors and mistakes. That method is to do *nothing*. In fact, an organization that rewards 100 percent error-free behavior will often have a lot of people who do little to nothing. This pursuit of the rules-based, error-free company will lead to the evolution of a company that does nothing. Consider this—if rules are abundant and mistakes are punished, then the only way to avoid breaking a rule is to do nothing. Doing something would require knowing all the rules and carefully avoiding making mistakes or inadvertently breaking a rule. Doing nothing is far more efficient than trying to walk the minefield of bureaucratic rules and regulations.

On the other hand, a dynamic entrepreneurial team feels obligated always to be doing something. Whether making improvements to existing plans or products, or changing the order of things, entrepreneurial teams are looking to do more. This concept is at the heart of complexity because complexity is doing something. In many cases, complexity means not only doing something but continuing to do something (failures included) until the current best road is found. Innovation is about complexity. Just as complexity hopes to explain the movement of electrons in orbit, complexity is about harnessing the apparently random immediate needs of social systems (Byrne, 1998).

Looking beyond this situation, there are several other reasons why the entrepreneurial team will be more successful: they will better leverage social networking skills, the community at large, and the flexibility of open teams (Godin, 2009). Each of these aspects offers important strength to a small team that is not impeded by highly structured organizational rules.

Entrepreneurial teams will be able to embrace the growing strength of social networking. The social networking boom has been supported by the development of many different Web sites that allow people to connect to others of similar background or interests. These social teams will be able to reach out to others in order to embrace a greater web of people who could become important stakeholders of the team. These networked people can assist in the development, planning, launch, or even sales of whatever product or project is underway. Consider the strength of having a host of different individuals with varying backgrounds, all which have an interest in a project (Duarte & Snyder, 2006). These people can serve as anything from cheerleaders to beta testers.

Entrepreneurial teams will be able to create teams of fans, followers, and friends who can offer perspective to any project. Imagine the strength that a small team would have if each team member had a secret cabinet of wise advisors. Every aspect of the project and of the team can be improved by invisible advisors who appear to be available to the team.

Entrepreneurial teams will also be able to flex the local community to assist with the goals of the team. A social and gregarious team can leverage everyone around them to be of service to the team. It does not require a great commitment by anyone but the fact that they can reach out across boundaries to garner assistance from others will certainly give them an edge over a team that is tied to whatever meager resources they have available. More makes for positive improvements; even Napoleon recognized that victory would favor the heaviest battalions. Entrepreneurial teams will always have heavier battalions when compared to similar-sized rules-based teams.

Rules-based teams will be unable to leverage either a social network or the local community because they will be tied by rules to keep all information, whether important or not, a secret. Furthermore, rules-based teams will never ask for more help for fear of being replaced or supplanted by others. A rules-based organization discourages individuals from asking or giving help to others because doing so would be to proclaim that their team is either too inadequate or too lazy to achieve their appointed project. Rules-based teams will march to their certain doom rather than ask for help from others.

Entrepreneurial teams are considered open teams and rules-based teams are considered closed teams. The difference is simple: entrepreneurial teams are willing to open up to others while rules-based teams remain closed to others. This disclosure to others can help gain support for an entrepreneurial-based team. After all, which team would an outsider trust more—the team that is willing to disclose information about the project or the team that is purposefully secret? The open team certainly has the advantage of communication and trust building not only to the outside world, but to the team itself as well. If a team has a culture of openness, the entire team will benefit as everyone will understand that openness is the norm and secret behavior is discouraged (or outright punished).

Open teams will also be able to manage the team as replacement team members become easier to develop and implement. Consider if a member of a close team were to suddenly leave the team (a new job). That team member takes all the information of the project with them and then it becomes the challenge of the remaining team members to replace the missing team member. The problem is that the experience of the team member leaves and there is usually not enough documentation to make the transition seamless.

With open teams, since information is more fluid and available, the replacement of a team member is easier. It still might not be seamless, but it certainly will be easier as more information is available and others will be able to fill in any missing information (Hamel, 2007). This becomes a significant advantage as it allows for the efficient replacement of team members (even the team leader) with fewer problems.

Thus, entrepreneurial teams have significant advantages over rules-based teams. Perhaps in the time before the information revolution, where information was power, a disciplined rules based-team had an advantage, but now where communication is power, a flexible entrepreneurial-based team rules supreme.

ANALYSIS OF WHOLE FOODS MARKET™ AND KEY COMPLEXITY LEARNING

Whole Foods Market™ has managed to key into two important elements of complexity in their organization. Whole Foods Market™ has imparted the entire organization with entrepreneurial spirit and self-organization. An entrepreneurial spirit is not uniquely complex; however, when combined with self-organization it makes for a more powerful force in business. This means that the organization is not only able to learn and take advantage of passing opportunities, but it can also quickly align the organization in a manner to take advantage of local opportunities that might not be present anywhere else in the organization. Unlike other organizations where solid sameness is required, Whole Foods™ allows for individuals to respond to their own marketplace and to react to their customers individually.

This self-organization is important because it allows Whole Foods™ to be as flexible as a small local market. This is an important element because it allows the company to make local changes to make it part of the community. Since Whole Foods™ has some larger opportunities as a larger company, it can take those important elements and combine it with local flair. So bringing this back to complexity, complexity research has recognized that self-organization is what sets organizations apart from other linear systems and organization. Just as the human body can fight off an infection with its integrated response, an organization that is self-organizing can meet a kind of local competition quickly and decisively.

Complexity research has found that there are three qualities of self-organization that are distinctly complex: self-referencing, increased capacity, and interdependent organizing. Furthermore, the greater amount of these qualities, the more self-organized the system will be resulting in increased performance (Lichtenstein, 2000, p. 133).

One interesting observation is that the classic management reaction during turbulent economic times is to consolidate decision making (decisions require higher approval, slash budget at a high level), limit the flow of information (restrict financial information internally, hide from competition) and tighten managerial controls (increase approvals, reduce spending authority). In stark contrast, a complexity-based organization encourages managers and employees to remain in line with the principles of complexity by providing employees with the company vision, providing greater information, and decreasing controls so that decisions can be made more easily even at lower levels.

Giving this kind of power to the lowest level of the organization changes the organization from a hierarchy to forcing the organization to become adaptive (Lichtenstein, 2000, p. 139). The key here is accountability, however, because the organization must be accountable for these decisions at all levels. The organization must hold people accountable for their decisions so that the organization can make the best possible decisions in difficult times.

A factor that is hard to control when an organization is moving toward greater self organization, which is almost instinctive, is the impulse for project managers to want to micromanage the enterprise during difficult times. When things seem to be going wrong, the natural reaction is to want to take control and to hoard information in order to make every organization decision. This is a difficult impulse to contain as it seems like the natural order but the project manager must resist. What happens is that the decisions do not necessarily get better when a project manager micromanages, but the project manager feels in greater control. The organization train may actually be heading off the rails and down a cliff, but the impression is that the manager can cope better with this type of change because of the illusion of greater control.

Whole Foods™ clearly understands that its type of business model will come into contact with various local and national competitors. The more flexibility they can give the local branch to combat these challenges, the better off they will be. For example, Whole Foods™ founds that it was not in direct competition with the national chains with regard to produce but it was in competition with the local produce markets and stores. In some locations (Arizona), the local Whole Foods™ locations have been garnering relationships with local growers to affiliate themselves with these local markets so they can not only offer that locally produced and highly regarded produce but they can offer their list of other locally produced goods. In essence, Whole Foods™ understands that food trends are always changing, so the more that they can stay abreast of the changes in people's tastes, the longer they can remain ultra-competitive.

SUCCESS ANALYSIS

Given all the hype and discussion about Whole Foods™, its success comes from four essential elements that at their core are not as superficially related to complexity as one would think. When one looks deeper, one can find a sort of symbiotic relationship between the customers of Whole Foods™ and other important business elements. The three elements that stand out the most are community, service, and marketing. These four elements do not seem so different than any other listing of successful businesses in the world, but the application of these ideals will reveal that complexity resides not on the balance sheet or the annual report, but within the hearts of the people of the organization.

Community

When Whole Foods™ gets it right, it is all about the community (Stringer, 2009). It is about their people interacting with other people, who in turn become customers. Those customers in turn interact with other people who recommend the store, which then creates new customers. It is all about the people who interact with others. And it is not always the way that one would expect the interaction to work. The interactions are symbiotic and it is not only the interaction of the people, but the reaction of the people involved.

The first aspect of community recalls our friend Lorenz and his butterfly diagram. One interaction can make a difference. If one person goes to a store and finds interesting and high-quality products, he or she is likely to tell others in the community about the experience. This kind of chain-reaction soft selling can help give a location new business for generations. Imagine how powerful a positive interaction can be if it leads to a flood of new business. It is true that a single ripple in the water can cause a tidal wave somewhere down the line. Understanding that a community is an interactive social system and treating the system with care and respect will always yield benefits.

The second aspect is associating oneself with the community by supporting local events that are food exciting and interesting to people. For example, in the south Florida area alone, Whole Foods™ is a proud sponsor of two different but very interesting local events. Whole Foods™ is a sponsor of the Fairchild Tropical Botanical Gardens Annual Chocolate Event (now with coffee and tea)™, which is normally held in January. This annual event draws thousands of people from the local area and Whole Foods™ sponsors educational events as well as advertising their own brand. The event offers education about food and also showcases the beautiful Fairchild Gardens, where thousands of species of living plants are on display in an idyllic setting nestled in Coral Gables.

The other event that Whole Foods™ sponsors is the South Beach Food & Wine Festival™, held in February. This annual event draws thousands of people to the beach with hundreds of food, beverage, and related product booths; multiple stages for food and beverage demonstrations; and educational events. This food and beverage bonanza is not only for community but also provides a trade day for foodies and oenophiles alike.

Both of these events not only affiliate Whole Foods Market™ with the local community but also support the local community itself. The Fairchild Tropical Botanical Gardens event helps support the local gardens, while the south Beach Food & Wine Festival™ benefits Florida International University. It is these kinds of interesting and educational events that Whole Foods Market™ sponsors all over the nation in order to keep their brand in the minds of the community.

Service

People in the food and beverage industry always state that it is not always about having the best food, or the most exotic wine, or the most interesting tequila, but it is about the service. Diners may not return to a restaurant after a bad meal, but diners are more likely to trash a restaurant's good name in front of others when the service is bad. Nothing can ruin a business faster than bad service. Additionally, it is not always about good service, it is about exceeding expectations and doing the right thing.

Good service and doing the right thing is where Whole Foods™ does a great job. They have knowledgeable people who do the right things. They listen to their customers because those who interact with the customers are those who make the decisions. This is an important learning because nothing bothers a customer more than not feeling listened to.

Another aspect of this service is the internal service given to team members. When people are given the authority to do great things, they can achieve more. Also, this treatment goes within the organization as well because if people are accountable for their own results and the results of their group, they will perform to the highest possible standards.

Again, the highest possible standards are not just a complexity matter; this is a matter of making good decisions. Rather than have to compete only on price with other locations, Whole Foods™ understands that offering top-quality products will yield top-quality prices. Offering an assortment that is better than the competition is a way to prove those higher standards. In this regard, these are the reasons why the service is perceived as some of the best in their industry.

Marketing

As a group, project managers are some of the worst marketing people in society. A good project manager and a good project team will not complain or request anything other than the bare minimum and rarely, if ever, will they come with a problem without the solution already in hand. Marketing is a strange realm of untapped complexity. Most businesses understand that they need to have a clear marketing system and that "it takes money to make money," so for the most part marketing departments are the land of spending, fat and padded budgets, and most likely the place where impossible promises are made.

This may be true in some organizations, and it may be the perception of most project managers, but a skilled marketing department can leverage complexity on behalf of their organization more regularly than a project manager can. The reason is that complexity is about people. People do not always act rationally about purchases; in fact, many purchases are based upon emotion, trends, perception, logos, and probably a dozen or more logic-free reasons. Emotional purchases are so pervasive in society that stores love to stock the shelves at the checkout counter with "impulse buys." Most project managers fail to see why anyone would make an impulse buy because the buying process should be anything but impulsive. Yet emotional purchases tend to trump rational purchases in our daily lives.

These concepts lead us right to the door of Whole Foods™. Whole Foods™ understands that people buy for appeal, they buy brands that they can trust, and they purchase food that is at least perceived as wholesome. All of these matter to buyers; Whole Foods™ understands that price may be important, but it is not always the deciding factor (Hamel, 2007).

Whole Foods™ knows that there are two parts to marketing: the marketing that happens in the store and the marketing that happens outside of the store. The marketing that happens inside the store has already been addressed with service section and the marketing outside the store has been addressed in the community section. Still, there is more to the message than what is already seen and the plurality of the message makes the marketing more powerful than the individual message.

The marketing outside the store offers a message to consumers that Whole Foods™ cares about the community, about education, about their suppliers, and about their customers. Whole Foods Market™ has started a marketing campaign called The Whole Deal ™. The company is reaching out further to the community, which is already pinched for money, and is highlighting deals that it is offering. This also includes an element of offering coupons (not new to anyone but Whole Foods™) in order to attract coupon mavens. It will be interesting to see if reaching out more to a new clientele will work, but it will likely remain successful in most markets.

The marketing inside the store offers a message to consumers about the products that they buy. It can be as simple as advising consumers that origin of the produce they buy, or the value of the product they may consider purchasing. This marketing is coupled with the message going to the outside in order to create a holistic marketing that makes consumers feel good about their decision.

Consider that the combination of the message is what makes the message more powerful. In fact, it makes people want to shop there again because they find the entire experience educational, informative, and social. The holistic approach applies complexity because it recognizes the social system of human shopping and caters to the emotional and intellectual elements at the same time, thus creating a message that is greater than any individual repetitive message.

CHAPTER SUMMARY

Whole Foods Market™ is currently not the strongest company in the world but it offers a level of flexibility unmatched in its industry. The organization has ways of keeping its fans loyal to the brand, while still managing to be economical sustainable. Offering a bevy of marvelous items that are distinctive and wholesome in a setting that is straight out of a Norman Rockwell picture is a way to blend community, service, and holistic marketing. Complexity pervades the company in its philosophy, its people, and its products. Project managers should learn to adapt these kinds of ideas to their projects and keep these ideas throughout their career. Imitating those that know is one way to future success.

Projects and cultures maybe different, but teams and people are still interactive systems that need to be treated responsibly and with respect. Understanding the strengths of self-organizing small teams and applying those ideas can help make the team more efficient and can lead to considerable benefits over the course of a project. Remember that all of these elements will only work if there is a solid accountability structure coupled with a strong team and individual reward system. It is not enough to tell people that they must perform, but they must be held to the higher standard.

REFERENCES

Byrne, D. (1998). *Complexity theory and the social sciences: An introduction*. New York, NY: Routledge.

Collins, J. C., & Porras, J. I. (1997). *Built to last: Successful habits of visionary companies* (1st ed.). New York: HarperCollins.

Dahlgaard, S. P., Dahlgaard, J. J., & Edgeman, R. L. (1998, July). Core values: The precondition for business excellence. *Total Quality Management*.

Drucker, P. (2001). *Management challenges for the 21st century*. NY, NY: HarperBusiness.

Duarte, D. & Snyder, N. (2006). *Mastering virtual teams: Strategies, Tools, and Techniques That Succeed*. San Francisco: Jossey-Bass.

Godin, S. (2008). *Tribes*. NY, NY: Penguin Group.

Giblin, E. J., & Amusco, L. E. (1996). Putting meaning into corporate values. *Business Forum, 22*(1), 14–18.

Hamel, G. (2007). *The Future of Management*. Boston, MA: Harvard Business School Press.

Kasdon, L. (2007), March. Whole Foods™ goes small. *Fortune Small Business. Time, 17*(2).

Ledford, G. E., Wendenhof, J. R., & Strahley, J. T. (1995, Winter). Realizing a corporate philosophy. *Organizational Dynamics*.

Lichtenstein, B. B. (2000). Self-organized transitions: A pattern amid the chaos of transformative change. *The Academy of Management Executive, 14*(4), 128–141. (Document ID: 65264520).

Stringer, L. (2009). *The green workplace: Sustainable strategies that benefit employees, the environment, and the bottom line*. NY, NY: St. Martin Press.

U.S. Army & U.S. Marine Corp (Ed.) (2007). Counterinsurgency field manual: U.S. Army Field Manual No. 3–24, Marine Corp Warfighting Publication No. 3-33.5. Chicago, IL: The University of Chicago Press.

Whole Foods™ Web site, www.wholefoodsmarket.com.

Part IV

Create Successful Project Communities

Part IV will review leadership, teams, and change and how these key management issues apply to complexity. A project manager must already be versed in these topics but this section will give the project manager new perspective in which to apply key learning when handling these topics. Part IV begins with an explanation of a complexity-based organization's leadership. This will be reviewed in order to give the project manager a template for leadership with complexity. The section will then review different organizations from larger communities to micro-teams (tribes) in order to explain how the size of a group will mandate a different application of complexity. Part IV concludes with a review of change management and how complexity can assist in handling change.

Chapter 14

Leadership of Complexity-Driven Organizations

Figure 14.1 If stones are the followers, then the leader is the river that runs through them.

WHAT DEFINES AN ORGANIZATION RULED BY COMPLEXITY

An organization that is ruled by complexity has three important elements. First, the organization subscribes to the point of view that leadership, teams, the organization, business, the world, the universe—none of these are ruled by a mechanical or empirical model. Mathematical models may be important in some areas but such models do not dictate the actions of social systems. Second, the organization understands that everything is connected. Systems may not be connected by direct linear relationships, but there is always a connection between all human systems. Third, the organization takes an approach that is akin to servant leadership or stewardship. The organization recognizes that leaders are there to guide, assist, and support, not to monitor, enforce, and direct. Any organization that subscribes to all of these points of view is utilizing complexity.

First, organizational leaders must understand that there is more than one right way toward success. Too often leaders become locked into what has been successful in the past rather than trying to figure out what will be successful in the future. A leader must embrace the idea that one cannot step into the same stream twice. In other words, society, technology, and organizations are continually changing. There is no assurance that what worked in the past will work again in the future. The reality of social systems is that what has worked in the past will probably not work again. One is unlikely to find a situation that exactly mirrors something from the past. Of course, there will be elements that are similar to past situations, so the leader (and organization) must be cognizant of these convergences and act appropriately when they occur.

Too often, leaders with experience attempt to template their experience upon every new situation. Rather than learn what could be done, too often individuals do what has worked in the past. If one finds a new leader impresses their ideals upon any change rather than trying to learn and apply what is appropriate, then the organization is lacking the fortitude of complexity. If the leader learns and then applies a solution, then the organization is ruled by complexity. If there are pressures upon the leader to seek the best solution rather than apply a new solution, then the organization is truly embracing complexity. A culture of complexity should supersede the individual and the culture should be driving change.

Second, the organization must recognize that everything is connected. Just as the butterfly flapping in Singapore might cause a hurricane in Florida, there must be an understanding that there is a relationship between actions. Unlike linear thinking that holds that an action causes an equal and opposite reaction, complexity represents the elements of the unknown. Just as in organizations that apply complexity, such as Whole Foods, there is internal and external competition between like and unlike areas. For example, a linear market would compare meat sales and productivity only to other market locations' meat sales and productivity, but Whole Foods teams might compare their labor-per-hour yield to all different areas of the store (Hamel, 2007). A linear organization would not consider this a fair comparison and would not consider the information relevant. Yet, this is extremely important when business times are lean. This becomes a short coming of linear thinking, and this is where linear thinking fails to grasp new trends in business.

If an organization is always looking at only a narrow set of variables that have a clear one to one relationship, how can the organization see the big picture? Organizations need to examine all aspects of the business, the competition, the community, and the planet (Rigby, 2009). Understanding this delicate relationship is critical for a complexity-driven organization. In the past, organizations had the feeling they were observing the big picture by keeping an eye on the market and the competition.

The reality is far more complicated than ever thought imaginable before. In the end, everything matters to the organization. This seems like a huge generalization, but in the end there is an important perspective to understand about this kind of statement. Organizations need to pay attention to everything. The environment that made an organization successful might change in the future in such a way the organization might not be as successful in the future, or worse, the organization might cease to exist. If the organization is not paying attention to more than the market and the competition, then the organization might find itself on the way out faster than they realize.

Third, great importance has been placed upon transformational leaders, but two subset leadership types are also very important to complexity-driven organizations. For the top organizational leadership in a complexity-driven organization, one should find servant leadership or stewardship leadership. These leaders at the top will be transformational; one should find certain critical elements in order to classify something as a complexity-driven organization.

Just as in the first two points, the organization must be nonlinear and the organization must be examining everything, the leadership at the top must also have a broad view of everything. Both servant leadership and stewardship are similar in nature but slightly different in application. Each can be successful in a complexity-driven organization but the differences must be reviewed in order to see the different inherent advantages and disadvantages.

Servant leadership is defined as leaders who chose to serve the constituency rather than to lead (Wren, 1995, p. 22). Individuals who chose to serve the highest priorities of the followers typify servant leadership.

This kind of leadership's purpose is to meet the requirements of the individual followers. Gandhi is an example of a servant leader. The concept of the servant leader breaks the idea of leaders holding a position of power and authority over a group. The servant leader does not impose his or her will upon others; rather, the will of the group is imposed upon the leader to serve the cause of the many. Rather than have a leader with a vision, the servant leader is simply the vehicle that transports the ideas of the group.

"Stewardship is defined as the willingness to be accountable for the well-being of the larger organization by operating in service, rather than in control, of those around us" (Block, 1993). Stewardship is about choosing empowerment and responsibility, without the feeling of entitlement (Block, p. 35). Stewardship is the combination of service to the followers and empowerment for the followers. Stewardship combines these two powerful concepts in order to better achieve credibility for leaders. Individuals who practice service and empowerment are leaders who embrace stewardship.

The concept of stewardship can be seen in customer-driven individuals, such as Stew Leonard. Stew Leonard runs a highly successful supermarket that has been consistently profitable by exemplifying service. To put it simply, Stew Leonard has been making money by giving customers exactly what they asked for (Block, 1993, p. 11). Stewardship is not caretaking; stewardship is a burning desire to serve. According to Sergiovanni, stewardship integrates servant leadership and other qualities in order to create a "leader of leaders" (Sergiovanni, 1992, p. 139). Stewardship revolves around the concept of being of service to others, whether to customers, followers, or stakeholders, as the core ideology of leadership.

Both servant leadership and stewardship are conceptually unique, but there are many similar factors. As a matter of comparison, since stewardship heralds itself as integrating elements of servant leadership, elements of servant leadership will be compared with elements of stewardship. The styles of leadership associated with servant leadership are egalitarianism, prosocial behavior, and altruism (Choi, 1998). Each of these elements will then be reviewed to discuss how these styles compare to stewardship.

Egalitarianism revolves around the equality of behavior. Servant leader behavior creates equality between the leader and the followers. The leader accepts no more than the followers, and the leader ensures this equality by accepting the final reward last. Egalitarian leaders defer reward until all members can receive similar rewards. A billionaire such as Bill Gates, who gives away a quarter of his fortune, although charitable, is not an egalitarian leader. This type of philanthropy is generous and laudable; however, this gesture does not give all the other followers of the group reward equivalent to the billionaire's remaining fortune. Rather, a leader that distributes all of his or her wealth equally among followers would be defined as egalitarian.

Stewardship offers democracy as a tangible benefit of stewardship. Stewardship nurtures democratic, egalitarian behavior. The Bill of Rights, and the concept of equality before the law, serves as fundamental examples of egalitarianism. Stewardship strives to leverage this heritage and the beliefs of such in order to foster a sense of unity and partnership.

Stewardship's concept of service forms a community. All forms of leadership rest upon the fundamental concept that power is granted from others (Block, 1993, p. 42). In some cases, leaders may seize power unethically; however, a community still exists. In stewardship, much as in servant leadership, stewards chose to serve to assist others, rather than to meet their own personal needs.

Prosocial behavior is again another supportive activity related to servant leadership. This type of behavior is one of giving thanks to followers for their contributions as well as speaking favorably of followers' offerings. Again, the leader is giving of them; however, they are not sacrificing anything. The gift of thanks is certainly welcome, but it only plays a role in servant and stewardship leadership behavior.

Finally, there is the style of altruism. Altruism offers self-sacrifice as a characteristic; however, altruism focuses on the giving rather than the loss. Altruistic leaders may give away their wealth in order to achieve financial equality in the group; however, this is different than deferring wealth until all members can share in an equal salary or benefit. The distinction is slight, the importance is in intent. In altruism, the intent is to give away one's personal items in order to achieve a greater equality.

WHAT IS DIFFERENT ABOUT LEADERS IN COMPLEXITY-DRIVEN ORGANIZATIONS

Successful leaders from Caesar to Charles de Gaulle have utilized culture and philosophy to their advantage. True leaders understand the value of blending the culture of the time with their own distinct philosophy. Culture is a construct that contains all the norms of a tribe. Leaders must use this knowledge to first support their ideas and then eventually drive it as a platform to expand their sphere of influence. Philosophy exists as a knowledge base in which to increase understanding through reflection and introspection. Leaders impress their beliefs upon others through various intellectual and nonintellectual methods. Business culture in the U.S. is based upon free enterprise and has strong currents of the belief in the survival of the fittest.

Complexity-driven leaders understand that their role as a modern leader means they will have limited time and limited contact with everyone in the tribe. So, given that the leader will not always be available for the team, the stakeholders, or the community, there must be a way that a surrogate leader can be consulted. In the past, this was accomplished by having multiple layers of leadership and the project leader would delegate authority and responsibility to others. However, even this situation is limited and often more expensive than a project will allow. Hence, the most important aspect of complexity-driven leadership organizations is the utilization of a clear leadership philosophy that offers direction, guidance, and clarity to questions others might have about the project. Complexity-driven leaders understand the social system of individuals and realize there is not a direct relationship for critical social concepts, such as learning, knowledge, and understanding (Hass, 2009). Finally, complexity-driven leaders understand that social systems operate more like a snowball traveling down a mountain than a cause-and-effect relationship (Figure 14.2).

Leadership philosophy is based upon the leader's ability to go beyond the clever spreadsheet, the informative report, and market forecasts. New businesses are being launched daily and with these new business developments come bold new beliefs about employees, customers, and stakeholders. While business culture remains monolithic, the philosophy of the leader is pliable. A leader must learn to adapt to the culture of business

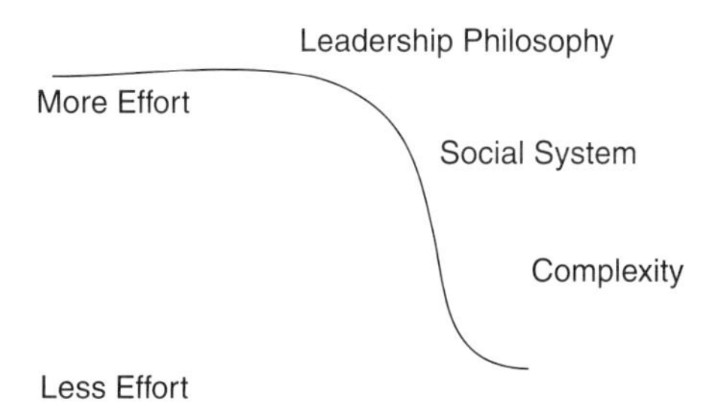

Figure 14.2 From leadership philosophy to complexity.

while instilling a successful philosophy upon the business venture. To better apply this to our own roles in leadership, we will review various business models of business culture and business philosophy in a comparative review format.

Business culture in the U.S. has more Darwinian attitudes than any other popular belief. Business is considered brutal and ruthless with the only rules being the laws of the jungle. Only the strong employees, customers, and stakeholders survive in this environment. This cultural belief helps drive the system to innovate, but the human price is high. Aggressively believing that businesses are struggling to survive perpetuates the myth that continuous competition in a free market is the only path to economic success.

"For Darwin, the metaphor was a branching tree of evolution, which he could trace the rise of various species" (Bennis & Goldsmith, 1997, p. 103).

Leaders in business culture today must understand the Darwinian attitude in order to learn how to adapt. Leaders must learn the "skill of adapting—being able to adjust or fit your behavior and your other resources to meet the contingencies of the situation" (Weiss, January 1999, p. 2). Leaders are examining the potential benefits from cooperative efforts in an attempt to enhance productivity. New forms of teams and the distribution of knowledge are being heralded as the innovative culture of the future. Leaders must remain informed about the business culture and utilize these norms to better perform their duties. Understanding the cultural forces that impact customers, employees, and stakeholders is critically important in business. Those who understand the needs of others and can adapt this knowledge in order to meet their needs will be the companies that become successful in the future.

A leader must have a philosophy that takes into account the currents within the greater business culture. A leader understands the larger culture of business, and shapes the business to interact with that culture. Another common virtue of so-called leaders is to avoid controversy and to follow orders. This runs contrary to the widely held opinion that leaders should be trailblazing individuals with a guiding vision. Unfortunately, when leaders allow others to deploy them, they are only serving to achieve mediocrity. "After all, most of us pass our time sheltered under the middle bulge in the bell curve. Too often people are huddled there in the mean of the curve of humanity, looking with envy and fear at the exceptional few pushing out the edges" (Hillman, 1996, p. 249). The trap of leadership drives many companies and businesses to failure. When leaders strive to mediocrity, everyone loses; stakeholders, customers, and employees all lose. Mediocre leaders do not motivate or inspire trust. Organizations that do not learn and grow will eventually wither and die.

In many complexity-driven organizations, there is often a propensity for servant-type leadership or stewardship. Since leadership takes on many shapes and forms, and complexity often lends itself to organizations that are more driven from the bottom, servant leadership and stewardship are more common. Both of these forms of leadership place the needs of the follower above the needs of the leader; however, these forms of leadership contain several key elements of differentiation. In order to better focus this comparison, it is necessary to first define each of these leadership concepts and then to discuss how these elements work with complexity.

Servant leadership is defined as leaders who chose to serve the constituency rather than to lead (Wren, 1995, p. 22). This idea is very much in alignment with complexity. Complexity is about harnessing the strengths of the tribe rather than trying to control the tribe. Servant leadership is more than a leader who puts the priorities of followers above his or her own. A true servant leader must also practice self-sacrifice. Individuals who practice self-sacrifice embody the deeper concept of servant leadership because the leader is physically and mentally enduring the same challenges as the followers.

Another similar leadership style that lends itself to complexity is stewardship. "Stewardship is defined as the willingness to be accountable for the well-being of the larger organization by operating in service, rather than in control, of those around us" (Block, 1993, p. 35). Stewardship is about choosing empowerment and responsibility, without the feeling of entitlement. (Block, 1993). Stewardship is the combination of service to the followers and empowerment for the followers. Stewardship combines these two powerful concepts in order to better achieve credibility for leaders. Individuals who practice service and empowerment are leaders that embrace stewardship.

Stewardship offers democracy as a tangible benefit of this type of leadership. Stewardship nurtures democratic, egalitarian behavior, which values equality and equity between all the members of the organization. Furthermore, since individuals are perceived with equality the culture becomes easier to maintain. Instead of the leader having to push the cultural ideas to all that will listen, the organization will naturally support the ideas.

The Bill of Rights, and the concept of equality before the law, serves as fundamental examples of egalitarianism. Stewardship strives to leverage this heritage and the beliefs of such in order to foster a sense of unity and partnership. Furthermore, stewardship tends more toward the concept of community as the driving social force within the organization. Stewardship is very close to complexity, but there are a few important differences to note.

Stewardship still aspires to the concept of hierarchy. The steward, often the CEO, is the head of the organization and remains the conduit for all these elements (Block, 1993). Of course, any animal must have a head and a heart; the concept of stewardship leaves that single leader as the singular individual who still runs the show. What differs with complexity is that it takes stewardship even further. Instead of the steward being a leader of leaders, complexity offers the concept that everyone is an important element and leader of the organization. Just as when Western leaders failed to understand Native Americans' lack of understanding of property, it seems that leaders are finally coming around to understand what Native Americans always understood.

The land is not for sale and cannot be owned by any one person because it is owned by all of us. The fundamental concept is often misunderstood; what is considered tangible property is never really a purchase but is always a loan. The earth is on loan to us for our lifetime and we in turn will pass what is left on to others, who will need to extend that loan. Complexity is about understanding; there is connectivity among all transactions. Just as the flight of a butterfly in Hawaii might create high winds in Florida, there cannot be a single steward for anything. In fact, each of us is a steward of many things. Each of us has a responsibility to our community, to our tribe, to our virtual team, and to one another. It certainly is easier to pull the sled when there is one lead dog, but each dog of the team must do its part. Each will lead, each will follow, and each will contribute to the movement of the sled.

To put it another way, if Rudolph with his nose so bright guided the sleigh on one foggy Christmas eve, what was Rudolph doing the rest of the time that put him in a position that he could actually be successful at leading the sleigh on that fateful night? Applying complexity hopes to explain what can be done the rest of the time in order to propel the virtual project manager to success. Leaders might be called upon to handle a situation, or they may not be called upon to handle that situation, but complexity offers more options than to intervene and solve, or to ignore.

Often the best a virtual leader can do is to support the tribe and to allow the tribe to resolve the problem (Godin, 2008).

LEADERSHIP ELEMENTS NECESSARY FOR COMPLEXITY

Leading with complexity is difficult in any organization, but in order to lead effectively while utilizing complexity, the project managers and the organization must be supportive of an environment that includes complexity. For complexity to be successful in an organization, it requires that there already be a system of solid leadership supported by the culture. Furthermore, the project manager must be an individual respected by the organization and considered the correct person for the job. Having a solid footing to start will assist in allowing the project manager to be successful. A leader must be made of the right stuff and be prepared to demand the best of stakeholders and those that are on the project team.

Leadership

According to Bass (1990), "Leaders facilitate interpersonal interaction and positive working relations; they promote structuring of the task and the work to be accomplished" (p. 383). Leaders are the positive relationship between themselves, their followers, and the work expected by the organization. Followers define the role of the leader just as much as the leader defines the tasks of the followers.

The role of the good leader is to navigate the uncharted waters of the future, while internally wrestling with the bureaucracy of the world. Leaders are those who move an organization forward by charting a course that others want to follow. According to Peters (1979), as cited in Bass, "top management's most important role is the shaping of the organization's values to provide coherence in an untidy world, where goal setting, option selection, and policy implementation hopelessly fuzz together" (Bass, 1990, p. 405). Kouzes describes leadership as motivating people to follow an individual willingly to a place they have never been (Kouzes, 1995). The role of a leader is more than overseeing organizational tasks. "Leadership is whatever discretionary actions are needed to solve the problems the group faces that are embedded in the larger system" (Bass, 1990, p. 386).

Five Elements of Good Leadership

In order to define the concept of leadership, it is necessary to review five elements of good leadership. Once these elements are understood and the leader is able to mirror these elements, the organization is prepared to move forward with complexity. To try to move an organization toward complexity without a solid understanding of leadership would be a waste of time. Without the leader being respected and the plan for complexity being understood, trying to move an organization toward complexity would be futile. At best, the leader might be able to develop small teams or pockets of complexity, or more likely the leader would make a mess of the organization by sending mixed and confusing messages to project team members and to stakeholders.

First, clarify the role of the leader to establish what makes someone a leader. Second, understand the difference between leadership and management. Third, develop an awareness of the relationship between the leader and the follower. Fourth, review the importance of the elements of leadership vision and leadership innovation. Fifth, discuss the importance of a personal leadership philosophy and cultural awareness. Leaders are those individuals who have strong beliefs and values that deal with the uncertainties associated with change. By reviewing these five areas, an individual will achieve a greater understanding of this elusive trait called leadership.

The Role of a Leader

A leader must have the ability to visualize the future and the capacity to serve (Fisher, 1991; O'Toole, 1996, pp. 34–35). Leadership is the ability of an individual to influence a group toward the achievement of a particular goal (Robbins, 1996). A leader should actively listen to what each individual has to contribute, and then counsel those who are in need of guidance (Fisher, 1991). Furthermore, leadership can be envisioned as a multilayered pyramid. The role of a leader begins with offering feedback and culminates with creating credibility. In between those layers is a thick layer of trust. Leaders should exhibit a balance between feedback, trust, and credibility.

Feedback is the organizational information between leaders and followers. "One needs organized information for feedback" (Drucker, 1996, p. 142). One must gather feedback from direct exposure to reality. Failure to connect this feedback with reality will result in poor decisions (Drucker, 1996, p. 142).

"Trust is the lubrication that makes it possible for organizations to work" (Bennis & Nanus, 1997, p. 41). Trust holds an organization together, and keeps the leader connected to followers. Organizations that operate from trust rather than hierarchical control are more effective, creative, and fun (Handy, 1997, p. 381). Trust is the reason why people work for a leader and continue to do so (Bennis, 1994, p. 160). Trust creates a bond between the leader and follower in a way that allows the follower to move toward an uncertain goal. Followers trust leaders who advise them something will happen, and then it does happen (Bardwick, 1996, p. 136).

Credibility is the ability of leaders to influence their constituency based upon their personal values and reputation. "Credibility is based upon six dimensions: conviction, character, care, courage, composure, and competence" (Bornstein & Smith, 1997, p. 283–284). According to Bornstein & Smith (1997):

- Conviction: the passion and commitment the person demonstrates toward his or her vision
- Character: consistent demonstration of integrity, honesty, respect, and trust
- Care: demonstration of concern for the personal and professional well-being of others
- Courage: willingness to stand up for one's beliefs, challenge others, admit mistakes, and change one's own behavior when necessary
- Composure: consistent display of appropriate emotional reactions, particularly in tough or crisis situations
- Competence: proficiency in hard skills, such as technical, functional, and content expertise skills, and soft skills, such as interpersonal, communication, tem, and organizational skills (pp. 283–284).

Leadership as Opposed to Management

Leaders create solutions, while managers solve problems. Leaders create a vision of a new reality for an organization, while managers are the implementers of the tasks necessary to create the vision. Managers operate the organization, and pay exclusive attention to bottom-line performance. In contrast, leaders are concerned with building the organization toward long-term success (Bennis & Nanus, 1997, p. 210). Leadership is working with people to accomplish goals, while management is telling people to work.

Leaders need an ability to look through a variety of lenses. We need to look through the lens of a follower. We need to look through the lenses of a new reality. We need to look through the lens of hard experience and failure.

We need to look through the lens of unfairness and mortality. We need to look hard at our future (De Pree, 1995, pp. 453–454).

The roles of both leaders and managers are important; however, powerful leadership is at the core of powerful organizations. "Management is about coping with complexity, while leadership, by contrast, is about coping with change" (Kotter, 1998, pp. 39–40). Leadership is never accidental; it must be planned and intended. "Leaders are made, not born" (Bennis & Goldsmith, 1997, p. 23).

Follower–Leader Relationship

The intention of a good leader is to create respect for the followers (Kouzes, 1995). The leader must find ways to raise followers from their current level to a higher level. A leader should intend to develop good followers; otherwise the leader is not motivating people to a higher purpose. Leaders who create good followers in turn create good future leaders. "Respect for the followers is made manifest by listening to them, faithfully representing them, pursuing their noblest aspirations, keeping promises made to them and never doing harm to them or to their cause" (O'Toole, 1996, p. 98). Leaders must always work toward the goal of creating respect. Respect comes from actions and intent; it is difficult to create respect accidentally.

Leadership Innovation

Leaders must focus on innovation, because innovation is long-term thinking, while management is about short-term thinking. Innovation is investing in a potential new process that will revolutionize an organization. "Innovative learning deals with emerging issues—issues that may be unique, so that there is no opportunity to learn by trial and error" (Bennis & Nanus, 1997, p. 181). Innovation is the creation of a new method, system, or process that is an order of magnitude more effective at achieving the same results than anything previously implemented (Peters, 1997). To embrace innovation is to embrace risk, because it is attempting something new. Leaders need to support more of the innovators in an organization in order to support this long-term approach. According to Schon,

"Radical innovations require champions that are committed, persistent, and courageous in advocating innovation" (Bass, 1990, p. 219). Recent real growth has come from the development of new solutions, which were significantly more efficient than their organizational predecessors. The victorious armies of the future will be those that can muster creative solutions in ambiguous circumstances (Cohen & Tichy, 1999).

Leadership Philosophy

Effective leaders have a personal leadership philosophy that is both internally understood and externally communicated. Leaders must learn to stand out in a crowd by understanding the business culture surrounding them while instilling a philosophy that supports their business goals and objectives. The acorn theory is an example of this philosophy of self-expression. The acorn theory states, "every person bears a uniqueness that asks to be lived and that is already present before it can be lived" (Hillman, 1996, p. 6). Hillman contests that this inner vision is the discovery of our daimon, which we have been given before we are born (Hillman, 1996, p. 8). Leaders must strive to discover their inner visions or inner voice, and have the courage to express these thoughts within their business organization. "You are your own raw material. When you know what you consist of and what you want to make of it, then you can invent yourself" (Bennis, 1994, p. 40).

If a leader can agree that all of these leadership elements are in place, then the leader should begin to move the organization toward the transformation to complexity.

CHAPTER SUMMARY

Leaders are those whom others will follow. And people follow those who embody the elements that they treasure within themselves. Managers are those who rule by authority vested upon them by others. There are many great types, styles and methods of leadership, but ultimately leaders are those whom others will follow from the known to the unknown. Consider your own type, style, and method of leadership and see how it can be improved, or better still, how one's unique leadership style can better leverage complexity.

REFERENCES

Bass, B. (1990). *Bass & Stodgill's handbook of leadership: Theory, research & managerial applications* (3rd ed.). New York, NY: The Free Press.

Bardwick, J. M. (1997). Peacetime management and wartime leadership. In F. Hesselbein, M. Goldsmith, & R. Beckhard (Eds.), *The organization of the future* (1st ed.), pp. 131–139. San Francisco: Jossey-Bass.

Bennis, W. (1994). *On becoming a leader*. New York: Addison-Wesley.

Bennis, W., & Goldsmith, J. (1997). *Learning to lead* (updated ed.). Reading, MA: Perseus Books Group.

Bennis, W., & Nanus, B. (1997). *Leaders: Strategies for taking charge* (2nd ed.). New York: HarperBusiness.

Block, P. (1993). *Stewardship: Choosing service over self-interest*. San Francisco: Berrett-Koehler.

Bornstein, S. M., & Smith, A. F. (1997). The puzzles of leadership. In F. Hesselbein, M. Goldsmith & R. Beckhard (Eds.), *The organization of the future* (1st ed.), pp. 281–292. San Francisco: Jossey-Bass.

Choi, Y., & Mai-Dalton, R. (1998). Leadership Quarterly: *On the leadership function of self-sacrifice*, Database: Academic Search Elite, UOP Online library.

Cohen, E., & Tichy, N. (1999, September). *Operation leadership*. Fast Company, Boston: Fast Company Media Group LLP.

De Pree, M. (1995). Leadership jazz. In T. Wren (Ed.), *The leader's companion: Insights on leadership through the ages* (pp. 453–455). New York: Free Press.

Drucker, P. (1996). *The effective executive*. New York, NY: HarperCollins.

Fisher, J. R. (1991). *Work without managers*. Tampa, FL: The Delta Group.

Godin, S. (2008). *Tribes*. NY, NY: Penguin Group.

Handy, C. (1997). Unimagined futures. In F. Hesselbein, M. Goldsmith & R. Beckhard (Eds.), *The organization of the future*, (pp.377–383). San Francisco: Jossey-Bass.

Hamel, G. (2007). *The Future of Management*. Boston, MA: Harvard Business School Press

Hass, K. (2009). *Managing complex projects: A new model*. Vienna, VA: Management Concepts

Hillman, J. (1996). *The soul's code: In search of character and calling*. New York: Warner Books.

Kotter, J. P. (1998). What leaders really do. In Harvard Business School Press (Ed.), *Harvard business review on leadership* (pp. 37–60). Boston: Harvard Business School Press.

Kouzes, J. M. (1995). *Achieving credibility*. New York: Simon & Schuster.

O'Toole, J. (1996). *Leading change: The argument for value based leadership*. New York: Jossey-Bass.

Peters, T. (1997). *The circle of innovation*. New York: Random House.

Rigby, D. (2009). *Winning in turbulence*. Boston: Harvard Business Press.

Robbins, S. P., (1996). *Organizational behavior: Concepts, controversies, applications* (7th ed.). Englewood Cliffs, NJ: Prentice Hall.

Sergiovanni, T. J. (1992). *Moral leadership: Getting to the heart of school improvement*. San Francisco, CA: Jossey-Bass.

Weiss, W. H. (January 1999). Leadership. *Supervision.*

Wren, J. T. (1995). *The leader's companion: Insights on leadership through the ages*. New York: Free Press.

Chapter 15

Communities

Figure 15.1 This cave is home to a community of bats. The only true flying mammals, bats live together in large groups known as colonies, which can grow to over a million bats. Bats are extremely social creatures and just like human societies the individual is connected to a family group as well as being connected to the community.

COMPLEXITY APPLIED TO GROUPS

Applying complexity to groups is not as difficult as many people believe. The hard part is not always the application but the identification of what can be done with existing project teams. This process has two distinct stages. The first stage of applying complexity to a group is the understanding of the interconnectivity between the individual and the group and between the individual and the community (Figure 15.2).

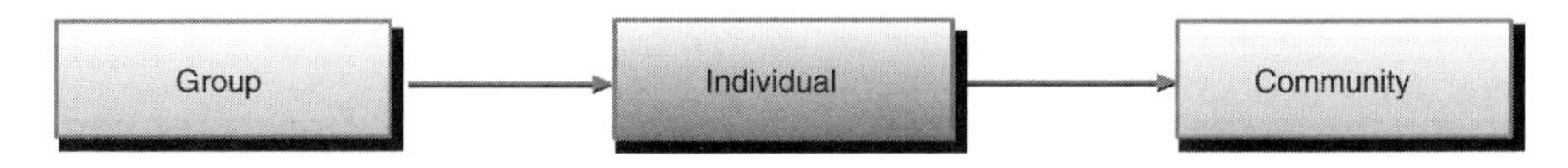

Figure 15.2 The relationship bridge: the individual links the group and the community.

In essence, think of the group as the bridge between the individual and the community. The second stage is when the project group systems are reviewed and streamlined. In the second stage, one must examine the project group to determine how complicated the internal systems are in order to determine if reducing the linear processes would be effective.

In the first stage, the project manager must make a concerted effort to recognize and map out the connections between the group and the individual and the group and the community. The project manager must examine these connections or "touch points." Determining these touch points requires asking this short list of questions about the individuals, the group, and the community.

Individual questions

1. Who does the individual communicate with?
2. Who does the individual trust?
3. Who does the individual go to with problems or questions?
4. Who is the official and unofficial leader for the individual?
5. Who is the individual's mentor?

Group questions

1. Who does the group communicate with?
2. Who does the group trust?
3. Where does the group go with problems or questions?
4. Who is the official and unofficial leader for the group?
5. Who is the advocate or champion for the group?

Community question

1. Who does the community communicate with?
2. Who does the community trust?
3. Where does the community go with problems or questions?
4. Who is the official or unofficial leader for the community?
5. Who are the fans or detractors of the community?

These may seem like simple and basic questions but each set of questions offers a different view of the organization. Without this basic understanding, it is impossible to apply greater complexity. In addition, this information allows individuals to better understand how they are connected to the community. It will also better connect the person to the project. Connections are important in complexity because when one starts taking away static linear systems and replacing them with flexible complexity based systems, people will naturally want to seek shelter in the known rather than have to face the unknown.

Once these touch points are understood, one can start to apply some best practices of complexity based groups. Complexity-based groups are effective by reducing processes, procedures, and other elements to allow the group to be highly flexible and nimble in business.

The group must also learn the discipline to know the difference between taking corrective measures and creating unnecessary systems. Complexity-based groups must remain nimble and responsive while being able to address the needs of the internal and external stakeholders of the organization. In addition, the project manager must understand the integration of the processes and must have a strong self-confidence of his or her ability as a leader (Hass, 2009).

Too often individuals feel that more controls are better and fewer controls are ineffective. The reality is that effective controls does not mean more controls; it just means that controls are enforced. Accountability is critical in complexity-based communities because there is little room to have people who are not adhering to the culture. The culture must be made to become a "bozo free zone" such as at Google (Hamel, 2007). The organization must systematically screen out anyone who is not carrying his or her weight as to avoid bringing in dead wood that would bring the organization down. Those who cannot meet the needs of the culture would be urged to move on in a hurry. This may seem harsh and Darwinistic, but it will keep the organization focused upon accountability and productivity rather than seniority.

COMPLEXITY APPLIED TO CREATE SUCCESSFUL COMMUNITIES

To understand how to establish communities with complexity, one must understand that communities are extremely complex and human history has been littered with scholars, scientists, and charlatans who have tried

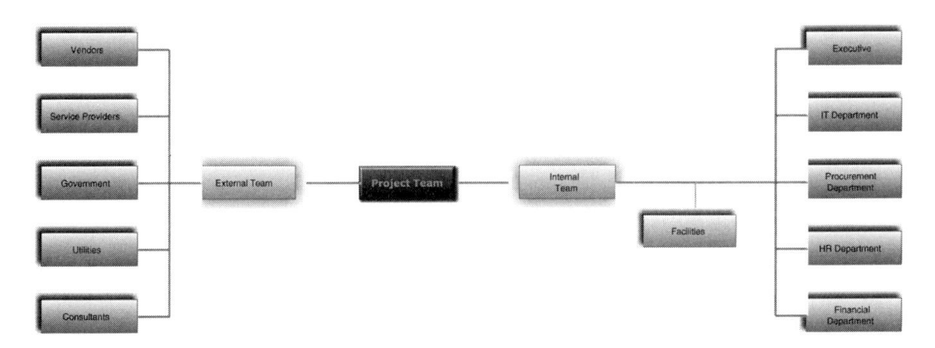

Figure 15.3 A sample project community.

to explain everything there is to know about society. Just glancing at Figure 15.3—which shows only the most common elements of project management society— conveys the idea that a community is highly complex. However, all communities are made of certain basic elements (Schein, 1992). All communities are based upon individuals, groups, (typically these are family units) and neighborhoods. All three of these elements end up with one thing in common: information.

Consider how communication might flow in the diagram shown in Figure 15.3. After some consideration, Figure 15.4 would appear to be the likely solution to the communication quandary.

The important consideration in this information flow diagram is that the project team holds the key to communication regarding the project. They will have some two-way interaction with the internal and external team, but for the most part the project team is the hub of information about the project for the community (Lipnack & Stamps, 2000).

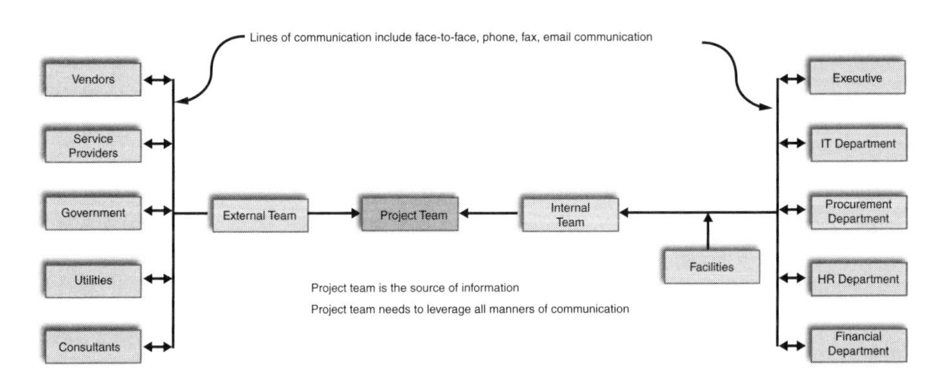

Figure 15.4 The flow of information in a sample project community.

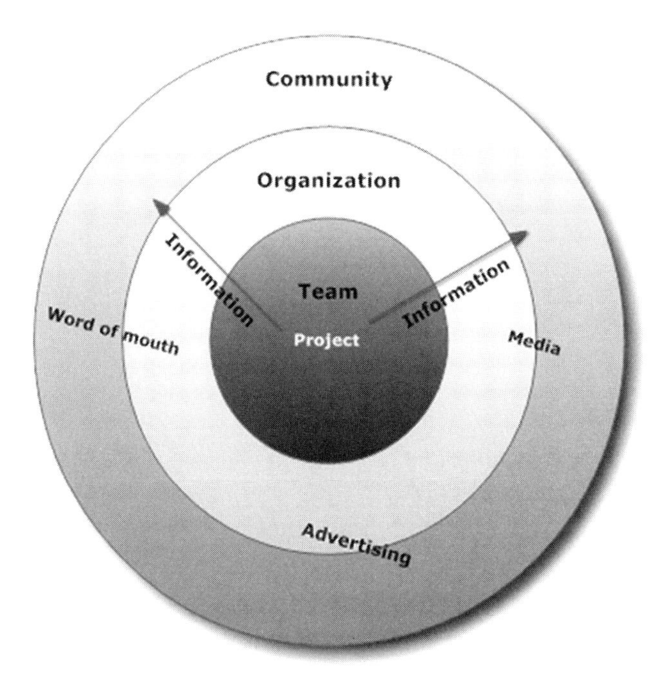

Figure 15.5 Project communication via complexity to the community.

This becomes the most important aspect of complexity that can be applied in a community.

This information becomes the kind of leverage that will assist a community be successful. A successful complexity-based community will harness the thirst for information about the project by making the information available to others on a regular basis (Drucker, 2001).

The secret to the success of the information is the fact that it must be done on a multimedia basis. The successful complexity-based community will offer information through various official and unofficial channels. Just as a new movie starts to gather attention—first from the hype created by the producers, second by the media, and third by word of mouth from viewers—a complexity-based community must do the same regarding a project. By offering multiple venues and outlets for information, the community can start to anticipate the project and create a buzz akin to a movie world premier. This is a method to leverage the social complexity of communities.

A typical complexity-based information delivery might proceed like this. First, the project team communicates the project and successes via email, newsletters, and formal business updates in a way to garner attention for the importance of the project. Second, the project team tries to leverage the media (no magazine or article is too small to consider) in order to better advise the community. Third, the project team helps stimulate the communication about the launch of the project in a way to help keep the idea in the minds of others.

Keep in mind that this communication should include all the stakeholders as well because outside contractors, vendors, and others want to be a part of a successful project, just as much as team members (Duarte & Snyder, 2006). The project manager must understand that stakeholders may require different types of communication. Leveraging the correct type of communication is important because a complex society wants to be a part of a greater purpose (Rigby, 2009).

The more communication and data that can be forwarded to as many sources as possible, the better social coverage the project will have. Communities love information and as one can see from society today, there is a never-ending desire for more information, particularly when that information is about something close to home.

CHAPTER SUMMARY

Complexity-based communities are more successful than other communities because these communities have reduced fractionalization, improved communication, and reduced complication. Communities that are able to come together more can be more productive and can allow individuals to feel more connected to the community. Many times this kind of community can be made when an external threat occurs, but in other cases they occur dynamically over time. Complexity-based communities are those that are connected in a manner where everyone simply knows who is responsible as well as what must be done. In some ways these complexity-based communities have an almost inherent vision that everyone understands.

REFERENCES

Drucker, P. (2001). *Management challenges for the 21st century*. NY, NY: HarperBusiness

Duarte, D., & Snyder, N. (2006). *Mastering virtual teams: Strategies, tools, and techniques that succeed* (3rd ed.). San Francisco: Jossey-Bass.

Hamel, G. (2007). *The Future of Management*. Boston, MA: Harvard Business School Press
Hass, K. (2009). *Managing complex projects: A new model*. Vienna, VA: Management Concepts
Lipnack. J., & Stamps, J. (2000). *Virtual teams: People working across boundaries with technology* (2nd ed.). NY, NY: John Wiley & Sons.
Rigby, D. (2009). *Winning in turbulence*. Boston: Harvard Business Press.
Schein, E. H. (1992). *Organizational culture and leadership* (2nd ed.). San Francisco: Jossey-Bass.

Chapter 16

Teams and Complexity

Figure 16.1 Urban sprawl is another reflection of complexity—each building is unique while they all fit together as a puzzle of humanity.

TEAMS AND COMPLEXITY

In order to understand how team and complexity are related, one must review the different types of teams that exist. Some of these team types are less common in project management today; having a complete understanding of the different types of teams is helpful to any project manager. In particular, this understanding will assist a project manager in determining what kinds of teams are available and may be in use. An organization may operate at a higher level as a dynamic and matrix-based organization (Cascio, 2000). There is still the possibility some teams or leaders have their own agendas and operate in a manner different from the rest of the organization (Bolman & Deal, 2003). Furthermore, if a veteran project manager is called to assist a team in trouble, that project manager must be able to quickly identify the types of teams in use to move them beyond their current formation. To this end, any project manager should carefully note the differences of these types of teams to avoid making an improper assessment that will cause team conflict in the future.

In general terms, teams are groups of people are tasked with a common purpose. Traditional work teams are groups whose membership spans multiple departments, such as engineering, finance, marketing, operations, purchasing, and sales. Multifunctional teams (multiple departments) can improve efficiencies by offering input from multiple sources. Consequently, traditional work teams are groups of individuals, with different skills and different organizational functions, who have come together to achieve a common goal while holding one another accountable for achieving these goals. Organizations have been using these types of teams or variations of these types of teams in order to improve organizational effectiveness. These types of teams are rigidly hierarchical and difficult to move toward complexity. Often it is easier to break up these teams and reform them than to try to change them. A project manager will find that these types of teams have the most resistance to change and are often more difficult to control than other types of teams (Lucey, 2008).

Linear work groups or silo teams are organized so each team member has an independent role and interacts with other team members on an ad hoc basis. Each team member is a specialist so it is generally not possible for one team member to replace another on the team. In this traditional formation, team members act independently and rarely interact, except to complete more complex tasks, and even then this interaction is brief and limited. For example, designers would complete the design process without any external feedback, and in turn they would pass along their product to research and development who build a prototype without any feedback from the designers.

This linear team process allows for departments to operate in their own silos without any interaction. The positive part of this kind of teamwork is that there is little reason for conflict. One the negative side, this kind of team is not efficient because the different departments struggle to change each other's final product. Had the teams worked together they could have designed a better product rather than have to build several iterations of the same product.

Again, this is a difficult team type to alter and can often cause problems when one is trying to transition this type of team to complexity. However, this type of team does have one advantage because it is the type that works best with the *PMBOK*® Lifecycle model. Rather than try to transition this type of team to complexity, one should utilize the traditional Lifecycle model from the *PMBOK*® to manage this type of team. This does not mean that complexity would not benefit such a team. However, the ensuing disruption would probably cause more problems with the project than would be acceptable. Rather than risk making organizational and cultural changes to this type of organization, one should follow the recommendations from the *PMBOK*® to achieve the best possible team.

A third type of team is the matrix team. Matrix team members have permanent roles but are often called upon to perform other functions. The efforts of the entire team ultimately facilitate the success of the project. This type of team is far more complicated because it requires that all team members communicate and have an awareness of the functions of all other team members. This type of team formation has roots in complexity. Teams that embrace complexity are those that can operate together seamlessly without continual hierarchical interference.

It is this kind of dynamic communication and awareness that will propel the team to success. The strength of the network will determine if the matrix team will be successful. A strong level of communication and acknowledgment is critical to keep the team from decaying (Roebuck, 2001). If the communication network is not maintained, people will feel isolated and duplicate effort will take place as team members start performing other team member's tasks. If all matrix team members keep their network advised regarding progress, it will keep the entire team moving forward. Another dynamic of this type of team is that acknowledgment, successes, and milestone achievement will propel the team forward. This type of team thrives on continuous movement toward the ultimate goal. More recognition and milestone achievement creates an atmosphere that will embody the higher purpose of the project.

Siemens created a successful kind of matrix team that crosses company boundaries and include individuals from different departments and different companies. When one division of Siemens was confronted by long lead times and increasing costs for the manufacturing of combustion turbines, Siemens chose to reorganize their teams to include external team members in order to increase productivity.

The Siemens matrix team consisted of a procurement specialist, a manufacturing engineer, a quality engineer, a design engineer, and a materials engineer. This team was partnered with a supplier's team, consisting of a process engineer, a quality engineer, a sales specialist, and a marketing specialist. Siemens has created a networked team in an effort to reduce costs, improve efficiencies, and increase communication. This team effort was assigned the acronym SMART, for S-Supplier; M-Management; A-And; R-Resource; T-Team. The team was established in such a way that each participant was a member group of equal partners, all of whom are autonomous, functional experts who participate together under the guidance of a speaker who is not the boss but is a transformational leader.

In the initial phase of the SMART team, role clarification removed the notion that these companies were in competition and there was also a period of time where the goals and objectives of the team was detailed so every team member would understand these goals. One key to making these multicompany teams successful was to migrate from a perception of organizational competitiveness (win-lose) to one of organizational cooperation (win-win). Since goals were created at the highest level possible, departments ceased to view each other as competition for the same internal resources but could then focus their attention on the bottom line of the project. This cooperative effort gives Siemens a mechanism to share more information with other companies involved in the manufacturing process. Other cooperative examples have a documented increase in productivity of 60 percent, as opposed to the competitive orientation, which cites 25–36 percent increases in efficiency (Krajewski & Ritzman, 1996). For Siemens, the results from utilizing nontraditional work teams were an overall success. Lead times decreased by 44 percent and costs dropped by 23 percent (personal communication with SMART team members, June 6–29, 1998). In addition, the payback period for Siemens for all cost associated with these changes was less than six months, with savings being organizationally visible in less than one year.

> Practical Tip: Teams can make a difference in an organization. No one is smarter than all of us.

Additionally, and now more commonly than before, matrix teams are also virtual. In order to clarify what is a virtual team the following definition has been amalgamated from several different experts in virtual teams. A modern virtual team is when more than 50 percent of the project team members are not resident in the same physical location, but are not necessarily dispersed over different time zones. The team depends on technology to communicate, rarely or never meets face to face more than once every two weeks as a project team, and team members themselves make decisions about the project. It is not unusual for one or two of the team members to rely on technology; however, it is more exceptional to have most members using technology to communicate with each other, the customer, and the project manager in order to accomplish objectives.

Most people believe that virtual teams are new; in fact, the concept of a virtual team is very old. The beginnings of an early virtual organization is seen with Moses in the Bible and the Roman and British Empires. In the Bible, Jethro chides Moses for not delegating day-to-day responsibilities to responsible men (Exodus 18:17–23, King James Version). Moses heeds the advice of his father-in-law and chooses "able men out of all Israel, and made them heads over the people" (Exodus 18:25, King James Version). Moses delegates his power to a few to enforce and keep the law of the land. This concept continues in history and we find many examples from the time of the rule of Caesar to the period of the British Empire. Communication is extended virtually through the placement of rulers in outlying lands and their use of human messengers. However, modern virtual organizations use technology to have almost instantaneous communication.

Networking technology, which includes information and communication, has created a new manner in which businesses communicate and has caused the exponential growth of virtual teams. The sophistication of technology is allowing companies to establish project management organizations that mirror traditional office PMOs. Studies of telecommuting programs or virtual programs indicate that a strong and stable relationship with a supervisor greatly increases the success of the telecommuting environment. Within a virtual PM environment, this stability is not the norm. Most projects consist of a team of individuals who may or may not know one another, and who are assigned to a project in a matrix management relationship.

Training and learning are major factors in the virtual environment's success. A virtual team appears to be more successful when training is conducted on communication skills and communication technology. Townsend and DeMarie's (1998) studies indicate that technology training should occur more often for virtual teams than for traditional teams since technology is the mainstay for communication and is evolving at a fast pace. There are also strong indications that the better the distribution network of knowledge and training in the organization, the better the team will function.

> Practical Tip: Consider investing in all the training that could possibly be necessary for any project team. The more training the team can receive, the more successful the team will be. Communication and technology are two forms of training that are always helpful to a project. Rarely does one hear a complaint that too much training is being done.

As a result of their study, Roberts, Kossek, and Ozeki (1998) find that executives dealing with virtual projects have three common issues: ensuring the correct skills are in the correct region/area when needed (¶13), disseminating innovative and "state of the art knowledge and practices" (¶14), and identifying the talent throughout the organization (¶15). English is the business language for all the companies within the study. However, this did hinder the virtual organization because of the different English grammar, English not being a native language, and the nuances of the various English versions. Roberts et al. (1998) also find that leaders in the eight companies are least adept at developing virtual solutions for teams. NASA was the most progressive institution with virtual solutions. In fact, virtual reality is used to train astronauts residing in countries other than the United States (Roberts et al., 1998).

Toney's (1999) benchmarking data further indicate that the superior project manager is professional in leading and managing; is competent in the technical field of the project; can articulate the vision to the project team; is constantly goal-oriented; and relates the goals to the organization. In addition, the project manager understands how to take advantage of opportunities and reviews alternatives (Toney, 1999).

The project manager's leadership style benefits from establishing an effective way to promote trust and collaboration in a faceless environment. Creative manners and opportunities help the project manager establish trust (Block, 1993; Toney, 1999; Duarte & Snyder, 2006; Jaafari, 2003). The effective project manager establishes trust between him/herself and each team member (Harshman & Harshman, 1999), and among the project team members. Studies indicate that without this trust, a virtual team is more likely to fail (Cascio, 2000; Hage & Powers, 1992; Jarvenpaa, Shaw & Staples, 2004; Kezsbom, 2000).

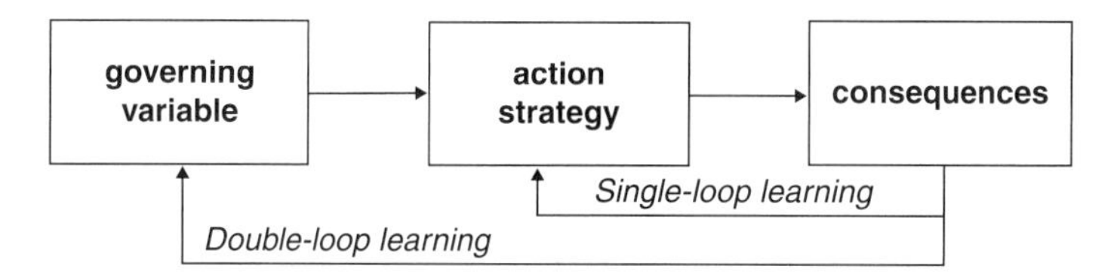

Figure 16.2 Single-loop versus double-loop learning.
Source: www.infed.org/thinkers/argyris.htm

An organization may institute change by implementing a double-loop method of learning, which could be advantageous in a virtual environment. This method takes advantage of the company's and the employee's knowledge and incorporates it back into the organization's processes and procedures. This method also can provide a catalyst for a streamlined, efficient, and ever-evolving organization that meets the demands of geographically dispersed clients, employees, and the company.

As illustrated in Figure 16.2, double-loop learning is important for virtual organizations because it ties actions and consequences to the governing variable. This makes the project (governing variable) the beneficiary of the learning. It may mean that a certain strategy may have negative consequences, but it does allow for the virtual project to benefit. For example, if certain quality shortcuts were taken and there was an increase in rework or rejections by the client, the negative financial consequences would impact the project. The negative consequences would in turn result in improvements or learning to avoid such quality problems in the future (Tichy, 2002).

> Practical Tip: Consider how double-loop learning can be applied to any prior project. Make the time to review a prior project to determine what can be learned from its success or failure. Great project managers learn from their mistakes instead of spending all their energy trying to bury their mistakes.

Boudreau, Loch, Robey, and Straud (1998) note that a virtual organization augments its chances of success by using a "federation concept" (¶13). The federation concept is described by Boudreau et al. (1998) as partnerships, joint ventures, consortia, and other creative alliances that change over time and with the needs of the virtual organization. This federation may include alliances with other organizations within the company or outside partners that may be required for success. This type of federation has been successful for the B-1 Bomber project, which had over 2000 corporations working together. Other successful corporations who employ the federation concept include Sun Microsystems, Nike, and Reebok (Boudreau et al., 1998).

The seamless integration of the technology within the organization and among the members allows local projects to have the support of a worldwide virtual organization, and the client does not realize that the product is a result of several companies or organizations. A well-run virtual organization should be able to function with very little regard to geographical distance and time barriers. A well-run and integrated virtual environment must be technologically seamless, responsive to local needs, and have the centralization necessary for efficiency.

Additionally, a federated virtual organization must be flexible and responsive to the needs of the environment in order to support the proliferation of partnerships and alliances. These partnerships and alliances will disband as needed, depending on the needs of the project and/or organization.

> Practical Tip: A virtual organization can be made to resemble a confederation of individuals who have been stitched together in order to operate together toward a common purpose. The more that the group is drawn together the more cohesive the bonds of the team. The leader should spend time bringing people together as well as communicating the goals of the project.

Application of Virtual Teams

One area of application of virtual teams is with organizational supply chains. Virtual teaming is responsible for the rapid collaboration of suppliers worldwide, thus creating an agile supply chain. According to Bal and Gundry (1999), a survey of the automobile (Rover) suppliers found the following virtual supply chain team benefits: 25 percent of the respondents indicated time savings, 23 percent of the respondents achieved cost savings, 20 percent realized an increased flexibility of resources, and 17 percent had a reduction in wasted time. Through examining the successful application of virtual supply chain teams in the automotive industry, it is believed that similar concepts can be applied with equal success in the maritime industry. Furthermore, Duarte and Snyder (2006) recognized virtual teams in successful organizations such as Hewlett-Packard, National Aeronautics and Space Administration (NASA), John Brown Engineers & Construction, Digital Equipment Company, and Rank Hovis (pp. 3–4). Virtual teams have implications in every organization, and the implications are not limited to large multinational firms.

> Practical Tip: There are benefits to working together. A team of people can be more successful than an individual in a game of tug of war. Teamwork has substantial benefits and the leader must keep in mind that a dysfunctional team will impede progress more than an individual bottleneck. Leaders must keep the team working together in order to achieve these efficiencies that have been previously described as part of successful teams.

Virtual Environment and Cultural Concerns

The virtual environment continues to evolve, and because of this evolution, there is great speculation about the future direction of virtual organizations. Kurzweil (2000) predicts exponential progress in science and engineering will allow humans to merge with machines. If this is correct, the human race will become resistant to diseases, think faster, live better, and become a species greater than what nature intended (Kurzweil, 2000). Along with these evolutionary changes, new organizations that can adapt to change will become even more important. Those who fail to adapt to the changes in the world will not only become unable to adapt but possibly become unable to interact with those individuals who have chosen to embrace technology and merge with machines (Kurzweil, 2000).

Many think the idea that humans will ever merge with machines seems like science fiction, but it does bring our attention to the enormous technological leaps that have occurred in recent years. In the last 100 years we have gone from the first powered human flight by airplane to missions to land men on the moon and space tourists going to the International Space Station. Regardless of where technology will take humanity, each of us must be open to new horizons.

Team Politics

Some organizations and managers are politically driven. Companies and human organizations are not mechanisms like cars or computers. Companies and human organizations are all about people. People can always choose to make a difference or to be a roadblock, depending upon their political disposition. Cogs in a machine have no such free will, and will always perform their assigned duty to the best of their ability, as long as they are not broken. Hence, one can expect either 100 percent productivity or 0 percent productivity from a cog. The cog either does the job or does not do the job. People simply do not operate the same way. People will offer productivity throughout the entire range of 0–100 percent and this productivity will depend upon a great number of variables. Unlike the cog that has two states, working and productive or broken and nonproductive, people have a wide range of productivity.

A talented manager must learn to navigate the political infrastructure of any organization. There is always an unwritten organizational chart that one must learn to utilize to advantage. While some individuals will have agendas, either hidden or apparent, the goal of the manager is to mobilize people to action.

> Practical Tip: Politics are always present in any organization. One must need to garner support where one can in order to avoid becoming a victim of politics. One can either accept that others will take advantage of their political influence and eventually erode your position or you can learn from these experiences and garner more far reaching support.

HOW TO ESTABLISH TEAMS WITH COMPLEXITY

To understand how to establish teams with complexity, there must first be an understanding of how and why individuals join teams. Many may believe that teams are formed when organizations force individuals to work together, this usually is not effective. Organizationally, placing people in close proximity with no incentive to work together and then to expect them to behave appropriately is not particularly effective. However, there are often reasons why individuals join teams and if one can understand those reasons, and then a project manager can leverage that knowledge along with complexity to develop teams with complexity.

A project manager might have some discretionary power to motivate teams with some financial incentives or other factors. If a project manager lacks that kind of leverage, the next best way would be to seek to understand why individuals have joined a team and then leverage that knowledge toward complexity.

On the surface, it would appear that the hardest part of the process is to identify why a person has joined a particular project team. Many managers are unsure of why certain team members have joined a team and that type of information does not always arise in casual conversation. A veteran project manager might already have some ideas on why an individual might have joined a team the next best way is to just ask. At first, the project manager might not feel that this is an appropriate question, but the reality is understanding; why people are part of a certain project can give a project manager valuable insight into why individuals are on a team. If the project manager knows what motivates an individual, this can help the project manager apply the correct knowledge and information to move that team member, and the organization toward complexity. To be truly effective, the project manager must be certain regarding the individual's motivation, so it is important to find out the exact reason why, because if the project manager is incorrect, the applied leverage might not be as effective. Furthermore, it will show the project manager that the answer will not be the same.

A project manager will find that not everyone on the team has the same focus or motivation. This will give some good information to start the process of moving the individual on the team toward complexity. It will also show the variety of reasons that people join teams and how different these reasons are. Given that there are many different reasons, it will assist the project manager in using a better method to move the team toward complexity.

ELEMENTS OF SUCCESSFUL TEAMS THAT LEVERAGE COMPLEXITY

Once it is understood why people join teams, then one needs to convert that information into a manner to move the team to complexity. To this end, there has been substantial research into the underlying reasons why people join teams. There tend to be five separate common reasons why people join teams. Because these reasons are driven more by the individual than by the medium of connectivity to the team, they apply to both virtual and proximate teams. Five of the most common reasons are security, status, self-esteem, affiliation, and power (Robbins, 1996). Each of these reasons is important in understanding the dynamics of any group. Understanding the underlying motive of a team member will help to understand the underlying human element of an individual within an organization.

The first human element of motivation toward team membership is security. Team members seeking security are those who often feel insecure or weaker when they are alone. A team member motivated by security may have concerns about real or perceived threats, and may want to have other resources available to them if challenged. "The insecurity of feeling isolated is replaced by security of a sense of belonging" (Bass, 1990, p. 656). Personalized management relations goes far to create a sense of security and belonging. Companies that treat employees as individuals and not like a number go a long way toward giving people a sense of purpose and belonging (Seidman & Skancke, 1990).

Individuals motivated by security are often those obsessed with the trappings of security. These individuals often look for the safest cars, the safest modes of transit, the safest places to visit. Individuals who find joining a team a safe way to be within an organization are often worried about loss. This could be a loss of lifestyle, a loss of financial stability, a loss of their job. The fear of all of these losses can drive a person to join a group. These are the kinds of individuals who often volunteer first for a new project; they are also the types of people who are already looking for the next project before the existing project is over. These individuals want to avoid ever being without a project because they fear that if they are not involved, they can be overlooked for promotion, or worse, that they can be replaced.

Individuals driven by security can often be leveraged by explaining the benefits of complexity to the project. By creating complexity as a form of security, it can assist in moving this type of individual toward complexity. Since complexity can assist in social interactions, this can be a point to explain to security-driven people toward complexity. Another aspect of complexity is that it explains, in the long term, how social systems work and this can often be interesting to security-driven individuals. If they can better understand how a project works, they can find this information to be useful in the future.

Because complexity is about social design, this kind of interaction creates team bonds that can benefit the individual beyond the project. This can also lead to greater security as the security-driven individuals see how this can apply to their next project.

The second human element of motivation toward team membership is status. Status can be important to team members if a particular group is considered with a very high regard. Bass (1990) states that those in positions of high status are perceived to have superior personal abilities even when those traits cannot be recognized (p. 176). The perception of superior personal abilities often attracts and retains team members to a particular group.

Individuals driven by status are often drawn to interesting projects and this draw can leverage them toward complexity. These types of individuals often are attracted to a project in the later stages, which allows these individuals to bask in the completion. These individuals are also easily driven toward embracing complexity because they want to show some level of accomplishment in the project.

A good assistance that these types of individuals can provide for the project is advertising. It is often a good idea to offer these individuals a social aspect of the project so they will communicate their accomplishment as well as the accomplishment of the team. An example of this would be to have them assist in the final delivery of the project as a task force member to get the final elements completed. The project manager should connect this final task to complexity by having these individuals recruit others to help at the end. An interesting aspect of people driven by status is that they often work and travel in groups of others with the same motivation. These individuals are able to draw upon nontraditional sources of assistance for the project. This moves the project to leverage a social network rather than traditional hierarchical elements. Those motivated by status can bring others to the project, to make the project deliverables more successful as well as to assist in the final stage of the project, when there is often a firm deadline that must be achieved.

The third human element of motivation toward team membership is self-esteem. Team members can achieve a sense of self-worth and esteem from being part of a larger cooperative. Bass confirms the importance and need for esteem (1990, p. 178). Often these team members are those who require constant support and reassurance from the leader that their contributions are valued. The esteem of leaders increases if the leaders effectively contribute to the team's success (Bass, 1990, p. 181). Individuals motivated by self-esteem are often also motivated by security because they are often also very insecure. So keep in mind that individuals are often motivated by more than one of these elements. If one can identify more than one motivating element, then the project manager should utilize that knowledge to magnify their use of complexity.

The fourth human element of motivation toward team membership is affiliation. Affiliation-motivated individuals are also often motivated by status; the difference is that affiliation-motivated team members are those who feel that the social networking aspects are more important than the status of the group. These individuals see that multiple group membership is more important than the status of the individual group. Group membership gives these individuals or groups the interaction that they feel is critical to their success. "Felt stress is likely to be reduced if an individual can be made to feel part of a larger entity" (Bass, 1990, p. 656). The sharing of information creates an atmosphere of trust and involvement, which in turn breeds affiliation. Involving all people at all levels creates a feeling of affiliation, which gives the impression that group members are part of a greater whole (Seidman & Skancke, 1990).

Affiliation-driven individuals can be motivated toward embracing complexity by explaining how complexity is the expression of group interaction. Complexity explains group interaction and explains how projects can leverage these networks. In many cases one can leverage these individuals' networks by including them on the project. An interesting aspect of these individuals is they are already applying complexity in their own lives, so it is only a matter of explaining complexity to them and they will begin to apply it with greater zeal. Once these types of individuals are aware of complexity, they will immediately move to explain this to everyone in their network to assist. It is often a good idea to include these types of individuals in any social event for the project as well as to include them in the planning of any kind of recognition event or project celebration. These individuals are often creative about getting groups of people together and are often connected with many different people, and more than likely they have ties to stakeholders in any organizational project.

The fifth human element of motivation toward team membership is power. Power-motivated team members are those who strive to motivate, influence, or control others. "Power motivated individuals are those that seek positions to influence people in different organizations and those in decision making positions" (Hughes, Ginnett, & Curphy, 1995, p. 347). These members drive the inherent value of a group in achieving goals that are elusive to the individual. The type of power described is socialized power, which is power exercised in the service of higher organizational goals, as opposed to selfish goals (Hughes, Ginnett & Curphy, 1995, p. 347).

Power-motivated individuals are often drawn to leverage complexity when they understand that transformational leadership is a relationship of power between individuals and it allows for a greater sphere of influence for leaders. These individuals can assist a project if they are put in a role of influence for the project. Often this type of individual would make a valuable champion for a project. By offering these types of individuals a role as a champion on the project, they can help spread the word about the project. When they gain a sense of positive affiliation and power, these individuals can become valuable advocates for the project. Power-driven individuals often have other affiliated motivations, so if one identifies someone motivated by power, it's a good idea to see if that person has other motivations as well. Power-driven individuals often have a method to achieve their power; some use affiliation and some use status, for example. In any event, as long as power-driven individuals can be convinced that complexity is a better method to power, these individuals can be key drivers in moving an organization toward utilizing complexity more.

In summary, there are many different approaches to motivating teams toward complexity and there are many different possible reasons to harness the potential of individuals. Ultimately, the goal of teams is simple: to energize people to performance levels above the expectations of the leaders (Katzenbach, 2000, p. 15). This peak team performance comes when the actions and the deeds of the organization are congruent; simply stating that people are the organization's most important asset without having the actions to back up their claims is not sufficient (Katzenbach, 2000, p. 201). Complexity becomes a successful vehicle toward making people the most important aspect of an organization by leveraging social networks and having leaders become transformational leaders. Teams can become the instrument of success or failure of an organization, and it is up to the leadership of the organization to determine if the organization's destiny is with or without complexity. Ultimately, the more people in the organization who are leveraging some type of complexity, the more the culture will move toward complexity. Often the process starts with a few footsteps in a certain direction, but over time and with encouragement, those few steps will transform into a stampede.

TOOLS TO MAKE COMPLEXITY PART OF ALL TEAMS

Complexity is about making more in a world of less. Few organizations can survive with a large research and development department that can monitor all of the needs of past, present, and future customers. Social systems are more about learning to work with the materials and items that one has. This is very uncommon in human systems, but is very common in the natural world. When we consider the diversity of plant, animal, and insect life on the planet and the large areas that these creatures inhabit, it is phenomenal that all of these species can manage to survive.

To take one example, consider the common garden slug. The slug lacks even the shell of its cousin the snail, and yet this creature continues to thrive. Slugs do not travel in packs or exist in herds where others are easily located, yet somehow the slow slug manages to survive despite lacking any kind of defensive shell, defensive speed, or any interesting camouflage. Slugs continue to exist and in fact thrive in the world. In fact, most gardeners at one time or another have to utilize pesticides or other measures to eliminate this problem pest. One must consider how such a species continues to exist when it has many known predators, is considered a pest by human standards, and is one of the slowest creatures in the animal kingdom. Even without the defensive colony structure of ants or bees, slugs manage to thrive in a world that really does not want them.

Other animals offer constructive metaphors for better applying complexity. Three animal types that can assist in understanding the application of complexity are the weasel, the fox, and the lemming. Each of these animals is associated with particular behaviors that lend themselves to complexity. These animals do not necessarily act with the knowledge of complexity; however, each of these creatures employs survival strategies that appear to be congruent with complexity.

Weasel

The weasel is typically associated with social cunning. In many ways, complexity is leveraging a social system in a manner that is most advantageous. A weasel might not be the strongest or fastest animal, but it knows how to utilize its environment to hunt and capture prey. The weasel stereotype may have some negative connotations, but it is a creature that understands the behavior others in a manner that helps it better survive. Thus the weasel is an individual expression of complexity in the way it uses its skills.

Complexity comes in the same form. Complexity is about leveraging the social structure of the team so that individuals can accomplish more. Some might think that this is a form of cheating, but in reality complexity is more about creating new value with fewer resources. This kind of frugal behavior is the type that should be harnessed during good times and bad. Project managers who can achieve greater accomplishments with smaller budgets will be those who will be successful in difficult economic times.

Fox

The fox is an animal most associated with being sly. To the uninitiated, utilizing complexity is considered being very sly. Organizations and individuals who are leveraging positive chaos or complexity are more in tune with one another. There is a high degree of trust, and with this trust comes individuals able to do more complicated tasks. If a task requires a complex set of steps and precision timing, then that would not be the best route for an organization lacking in trust. Such organizations must have clear guidelines and firm procedures to keep all the cogs of the system moving forward. Any deviation creates confusion and can disrupt the system. An organization with a number of highly autonomous employees who trust one another will be open to taking risks and trying new things on an ad hoc basis. These individuals understand that they are able to take these short-term risks in order to achieve greater results, while others would be less willing to take such risks.

These risks would be considered cunning or sly because others might see these risks as poorly conceived. However, since there is trust in the organization, these individuals understand that they will not have to face severe punishment for acting outside of their normal guidelines. So, if individuals on the project team are trusted and secure in their roles, they will be more willing to take risks that can benefit or improve the project. Since many benefits are very short lived, it may require that they take the initiative in order to reap the maximum benefits.

Lemming

The lemming is a creature best known for its extreme social behavior. Complexity is also about social behavior to the extreme. Although complexity-based teams are not likely to jump off a cliff together, they will be social and trusting to the point where they will follow their leaders to the extreme. A team that is trusting and understands that team members might not have all the information will be willing to follow their leader to a new place, without knowing or understanding the ultimate destination.

A project may require that individuals perform important tasks that are not immediately understood as part of the larger picture, but individuals who are leveraging complexity will have the trust that is necessary to achieve what is necessary. Just as troops follow their commanders into conflict situations, complexity allows for teams to trust one another enough to follow one another into difficult situations, even into situations that are apparently highly stressful or otherwise difficult.

Thus when one feels comfortable enough with complexity as part of a project, one should consider which animal(s) they are utilizing at which point in a project. Each animal may not be appropriate for every task of a project, but it is best to understand which one is in use in order to understand what elements of complexity are being used.

CASE STUDY: ADDING COMPLEXITY TO A MATURING COMPANY

Teamwork at Alpha Engineering

Alpha Engineering (AE) had been a successful and profitable business for a dozen years. The existing engineering group continued to increase sales by double digits every year for the past ten years. They mostly increased business by their stellar reputation, but the recent economy has slowed down considerably. Alpha Engineering has two offices, one in Los Angeles and one in Miami. The Los Angeles office is loosely referred to as the headquarters office.

New building projects have become scarce in a difficult economy and now AE faces their first year's loss in over a decade. In addition, AE has few projects on the books for the following year and now everyone is concerned that the company might not exist by next year. This has led the board of directors to carefully examine what has changed in the company to cause this catastrophic situation.

Due to the future of the company being in jeopardy, the board of directors commissioned an external consultant to audit the organization and found that the largest internal element of risk to the company was the profound lack of teamwork. The consultant found that employees had their set roles but no one made any efforts to improve efficiencies, and there was no effort to utilize past learning to improve the organization. This was found to be a high-level risk factor that could lead to the failure of the company during the current economic slowdown.

Part of this feeling stemmed from the fact that the current management believed each project was unique and hence they would develop new tools and processes for each project. The culture supported this feeling of uniqueness, even when there were similarities in projects. What the consultant uncovered was that 80 percent of the projects were intrinsically similar in process and procedure and 20 percent of the projects in the last two years were 98 percent identical to something done at AE in the last five years. Management had always been careful to allow for complete autonomy of project managers, but now this culture was causing the company to lose its competitive edge.

The consultant found that AE had six pending proposals to clients (five out of the Los Angeles office and one out of the Miami office). It was felt that if AE did not secure most of these projects, the organization would have difficulties at the beginning of the next year. The consultant recommended that an internal team be developed to work on the proposals so they could apply existing project plans, details, and processes to their upcoming bids. If AE were able to use prior material and apply its organizational learning to the upcoming projects, then the company should be able to offer extremely competitive bids because they would have a much shorter time for completion and a much lower cost in the development stage. The problem was that there were only 45 days to apply the best practices and to complete the scope of these bids. Most project managers like to have 30 days to complete a proposal, but in some cases they have been able to complete them in 15 days or less. Furthermore, someone needs to build the team and to lead them to success. The consultant recommended that a team be made up of some of the senior project managers, but they did not offer a proposal for the composition of the team as they felt that someone at AE should decide the best approach.

The following people were recommended by the consultant to be part of the team.

John Masters. John has been with the company for the last eight years and has been involved with many different projects. John specializes in the big, expensive projects and is often accused of spending too much, but he always delivers big results to the client. He is also effective at meeting a tight deadline. The largest complaint that people have about him is that he is a ruthless task master and lacks a soft people side. He tends to dominate a conversation and he can push people to get top results. He is also sometimes accused of being too picky about the details of a project, but no one can dispute his results. He is focused on two projects that will be finishing up in the next few months. John moved from the Los Angeles office about two years ago and now works out of the office in Miami, but is currently on site in Orlando.

Jacob Smalls. Jacob has been with the company only two years, but he has been involved with a few high-profile projects. He tends to lead by example and puts in a lot of hours for his projects. People find him easy to work for, but sometimes people take advantage of his kindness and understanding. His projects tend to be very good and on budget, but sometimes he misses a milestone deadline. Some people criticize him for doing too much and not delegating enough, but he feels that quality often can only be achieved by doing something yourself. He is proficient at completing proposals quickly and has been able to turn around some proposals in a few days. He is just finishing up a project and will be available in the next 14 days. He has always been in the Los Angeles office.

Cassandra Cleary. Cassandra has been with the company six years and has progressively moved up until she has attained the role of project manager. She started working in administration and spent some time with contracts, and she is proud of her achievements in project management. People like working for Cassandra and she leads by consensus. She always garners feedback from her team and implements as much of it as possible. She often sides with her team rather than trying to meet all the requirements of the client, but most people find her work to be good and she has a good reputation in the company. Her budgets are always met and she does very good work in determining all of the details necessary for the project. She also has a good understanding of contingencies and she always seems to have a plan B if the initial approach does not work. She is available now and she is currently in the Los Angeles office.

Sam Jackson. Sam has been with the company four years and has always been considered a solid producer. His projects tend to be the ones that are direct and smaller in nature. He is known to negotiate for every penny and does very well at showing a profit for the company, even when the project is not considered to be a very profitable project. He is known to scrounge resources from all over and he tends to reuse a lot of resources. He always makes budget, he sometimes loses sight of deadlines because he gets too involved in a negotiation. People like working with him because he manages to get bigger bonuses for his team due to the profitability of his projects, but they find that he does delegate a lot of the details of a project. He will be available in a few days as his last project is completed and he will return from vacation. He is currently in the Miami office.

Fred Timber. Fred is new to the company (less than one year), but he brings a wealth of knowledge from different competitors to Alpha. Fred has worked for three of the companies that will also be bidding on the projects for next year. Fred has been a project manager with all of those companies and has been successful at all three. He came to Alpha because he felt that he would be treated well with the company and they are located close to his home. He has a good reputation as a solid producer and his projects tend to be on budget and on time. Few people have worked directly with him on projects with Alpha; most people have heard of him working for the competition. Fred is currently in the Miami office and is available to start on any project immediately.

Scenario Questions

- How can complexity assist with the upcoming bids?
- How can complexity tools assist with improving the culture of the company?
- Who should be appointed as team leader?
- Who would determine the best practices of AE and to build the upcoming bids?
- How can the butterfly effect be applied in this case, given that many of the projects have been related?
- Who would be the least risky individuals to put together on the team?
- What kind of politics might happen with the interaction of the team? How can this be mitigated?
- What obstacles might be encountered by the team as they try to implement their findings? Why?

CASE STUDY REVIEW

There is a degree of team building that must be done with this group, and furthermore there is an immediate need for creating greater trust and understanding within the group. The case study is one of needing to create motivation within less than motivated individuals. Although these team members have their individual strengths, there is little reason to believe that they will work well together without strong motivation. The secret to success with this group is motivation. It is important to harness two elements in this case study. First, as the project manager one needs to offer highly coveted rewards in exchange for cooperation. Second, the project manager must understand what motivates each of the individuals.

The first element can be achieved by offering something lucrative for individual and team effort. It is important to offer incentives for both because if you only offer one, you will only achieve half of what is necessary. The second element is to learn which of the five most common reasons for team participation are most motivating for these key individuals. It may take some time to determine if security, status, self-esteem, affiliation, or power best motivates a person, but once one understands which element is motivating, the element rarely changes.

Thus, to best leverage a team one must first offer an incentive (positive is often better than negative) for the group and the individual and understand what motivates the person. Then one must convert this knowledge into something that is both interesting and motivating in order to keep the team on track.

CHAPTER SUMMARY

Virtual teams are now part of the business landscape and practically all large projects have some element of virtuality. A project manager must learn to harness the strengths of a virtual project, in order to be better off in the future. Projects will only get more and more complex and there is no doubt that project teams will in turn become more complex. This is where a project team leader can utilize complexity to make them successful. By understanding that transformational leaders are better equipped to handle complex projects and complexity, a project manager can adopt some of these skills in order to better organize virtual teams. The issue is often one of application rather than knowledge. Most managers know what the right thing to say is, but it is hard for a manager to do the right thing. Also, a project manager must learn to cope with conflict in order to remain successful.

Virtual project managers must learn to leverage the tools of complexity in order to be more successful in the future. This is not only managing conflict, but it means being able to manage a project with high expectations.

A project leader must learn to motivate others by offering a clear goal and clear expectations. The more the leader and the team can manage stretch goals, the better off the project will be. Greatness comes from doing great things with fewer resources than others would expect.

The military axiom of victory follows the army with the heaviest battalions is the expectation. So when a military leader achieves victory with a substantially inferior force than an opponent, that leader is often admired. When victory is achieved by a leader when defeat is expected on a consistent basis, then that leader is considered brilliant. Many times, that brilliant leader's techniques and processes are studied so that future military leaders can learn from their success. Greatness is often studied in order for others to replicate those successes in the future. Virtual project managers must learn to study the great leaders from the past in order to apply these strategies in the future.

REFERENCES

Bal, J. & Gundry, J. (1999, October). Virtual teaming in the automotive supply chain. *Team Performance Management: An International Journal, 5,6*, pp. 174–193.

Bolman, L. G. & Deal, T. E. (2003). *Reframing organizations: Artistry, choice and leadership*. (3rd ed.). San Francisco: Jossey-Bass.

Boudreau, M., Loch, K., Robey, D., & Straud, D. (1998). Going global: Using information technology to advance the competitiveness of the virtual transnational organization. *Academy of Management Executive, 12*(4), 120.

Cascio, W. (2000). Managing a virtual workplace. *The Academy of management Executive*, *14*(3), 81.

Duarte, D., & Snyder, N. (2006). *Mastering virtual teams: Strategies, tools, and techniques that succeed* (3rd ed.). San Francisco: Jossey-Bass.

Hage, J., & Powers, C. (1992). *The post industrial lives: Roles and relationships in the 21st century*. New York: Sage.

Harshman, E., & Harshman, C. (1999). Communicating with employees: Building on an ethical foundation. *Journal of Business Ethics*, *19*(1), 3. Retrieved January 28, 2001, from ProQuest.

Hughes, R. L., Ginnett, R. C., Curphy, G. J. (1995). Power, influence, and influence tactics. In T. Wren (Ed.), The leader's companion: Insights on leadership through the ages (pp. 339–351). New York: Free Press.

Jaafari, A. (2003). Project management in the age of complexity and change. *Project Management Journal, 34*(4), 47–57.

Jarvenpaa, S., Shaw, T., & Staples, S. (2004). Toward contextualized theories of trust: The role of trust in global virtual teams. *Information Systems Research, 15*(3), 250.

Kezsbom, D. (2000). Creating teamwork in virtual teams. *Publication of the American Association of Cost Engineers*, 42(10), 30. Retrieved November 27, 2000, from ProQuest.

Krajewski, L. & Ritzman, L., (2001). *Operations management, strategy and analysis* (6th ed.). Reading, MA: Prentice Hall.

Kurzweil, R. (1999). *The Age of Spiritual Machines*. New York: Viking.

Lucey, J. (2008, winter). Why is the failure rate for organizational change so high? *Management Services, 52*(4), 10–18.

Project Management Institute (Ed.). (2008). *A Guide to the Project Management Body of Knowledge—Fourth Edition*. Newtown Square, PA: PMI.

Robbins, S. P. (1996). *Organizational Behavior: Concepts, Controversies, Applications* (7th ed.). Englewood Cliffs, NJ: Prentice Hall.

Roberts, K., Kossek, E., & Ozeki, C. (1998). Managing the global workforce: Challenges and strategies. *Academy of Management Executive, 12*(4), 93. Retrieved June 17, 2002, from Academy of Management.

Roebuck, D. (2001). *Improving business communication skills* (3rd ed.). Boston: Prentice Hall.

Seidman, L. & Skancke, S. (1990). *Productivity: The American Advantage* (2nd ed.). New York: Simon and Schuster.

Tichy, N. (2004). *The cycle of leadership: How great leaders teach their companies to win*. New York: HarperBusiness.

Toney, F. (1999). *Competencies of superior project managers in large functional organizations* (*Volume I*). New York: Marcel Dekker.

Toney, F. (2002). *The superior project organization: Global competency standards and best practices*. New York: Marcel Dekker.

Townsend, A., & Demarie, S. (1998). Keys to effective virtual global teams. *Academy of Management Executive*, 12(3), 17. Retrieved March 5, 2002, from Academy of Management.

Chapter 17

Micro-Teams (Tribes)

Figure 17.1 Elephants are highly social animals and the females remain together as an extended family group. This kind of team is important to bond individuals in a form of tribal kinship.

According to the Counterinsurgency manual of the U.S. Army and Marine Corps, a tribe is an autonomous, genealogically structured group. The rights of the individuals in a tribe are determined by genetic relationships and connections to particular lineages. Furthermore, tribes are considered adaptive networks organized by kinship that are organized based upon physical and economic security. Many of these points are important for a project-related tribe. In many cases tribes hold stronger bonds when they are bonds of kinship, but these bonds are not required for a tribe in a project sense (U.S. Army and Marine Corps Counterinsurgency Field Manual, p. 86).

Tribes, in the project sense, are a group of people brought together for a similar purpose and focused upon a goal. The tribe must communicate internally and externally and must be led; otherwise they are just a crowd or, worse, an unruly mob. The ties among individuals need to be such that they make the group stronger toward the goals of mutual economic and social security. Essentially, tribes are small groups tied together by strong bonds that go beyond the typical kind of relationship that one finds in a sterile business environment.

Tribes are based upon the concept of mutual support (Merriman, Schmidt, Ross, & Dunlap-Hinkler, 2004). In the military this concept is necessary because if the enemy is well supported they will be harder to defeat. From a military standpoint, an entrenched foe that is determined and fighting for more than national pride and following orders will be a more difficult foe to defeat. To extend this beyond the military definition, a project team with more than a paycheck at stake will be more diligent and such a group will be less likely to fail. Understanding this concept can offer the project leader something to consider when the project runs into trouble. If the project team is invested in some way in the project and there are micro-teams present, there is a greater probability that new solutions will be found. Often, ideas spawn from necessity and necessity comes when others are counting on a person rather than just a paycheck.

THE IMPORTANCE OF SMALL TEAMS

There is a dearth of information about teams and how they are important in business. One cannot read a text on leadership, management, or business without having at least a small section dedicated to teams. Any project manager knows they must harness enthusiasm, skills, and abilities of the team to focus upon the project in order to achieve success. Even the *PMBOK® Guide* (PMI, 2008) writes about teams and the importance of them; however, one needs more direction on such a broad topic. Since it is impossible to create a definitive text for a broad topic, such as teams, strong guidelines are important in order to move a group toward success. By offering material on micro-teams, as well as the complementary concept known as tribes, this text can help guide the reader toward two of the more important recent complementary concepts in teams.

A micro-team is a small group of people (usually two or three) brought together for a specific purpose who communicate among themselves and have some type of leadership structure. Sometimes mundane micro-teams focus upon a simple goal and other times it is something grander. This concept is not new; it does have some powerful and very social connotations in society. A common micro-team in society today is the couple. In some cases there are formal relationships among couples, such as marriage, but in other cases there are less formal relationships, such as when two people are dating. The relationship of a couple seems very natural and normal within society and in many situations.

When one examines the structure of a couple in a relationship, one finds that all three of the elements exist: communication, purpose, and leadership. In a relationship there is—or at least should be—communication between the individuals. Couples regularly communicate and they will communicate regarding a variety of topics. This kind of communication is very important and often when one examines successful couples, one will find that one of the important elements of this kind of relationship success is based upon communication. Communication within the micro-team offers individuals a sounding board, an advisor as well as the potential for the development of new ideas (Godin, 2008). Not all of this kind of communication is constructive, but it is important in the project process. A micro-team operating effectively will be able to problem solve, review issues, and develop action plans without being directed to take action. This is a very powerful force, but part of this success is due to the communication.

In fact, society recognizes that communication between legally sanctioned micro-teams (married couples) is important to the relationship and is not always for public scrutiny. Society is very protective of this kind of intimate and private communication and according to common law, spousal communication is as privileged and confidential as between an individual and a lawyer, a doctor, or a member of the clergy. Hence this type of communication is not only private but is considered important and this kind of privacy is supported by the court system (Merriman, Schmidt, Ross, & Dunlap-Hinkler, 2004).

Bringing this back to project micro-teams, communication between micro-team members is not legally private (unless you are married to one of your project team members); however, this kind of privacy is important for the team. This does not mean that the team should actively subvert others or attempt to deceive others by miscommunication, but it does mean that part of the micro-team unwritten charter should be that communication is considered private and should not be repeated to others without permission. Of course, if this is taken to the extreme, one may start to believe that micro-teams are a kind of secret society that will result in some type of dark cult. The reality is far less mysterious.

It comes back to a simple understanding that children grasp better than the typical business professional does. Growing up, most of us had a best friend. In many cases, some of us still have that same best friend, or perhaps a new best friend. Consider micro-teams are working with your best friend. A person should not willingly embarrass or betray a best friend, and the material one discusses with a best friend should not be divulged to others. This kind of privacy and respect of communication will go a long way to making a successful micro-team. Respect and trust is hard to earn and always hard to get back once lost. Better to build trust and keep it than to constantly be trying to rebuild it after betrayal. Essentially, trust, purpose, and equality are the three concepts important to successful micro-teams.

Trust is the foundation of any team, and micro-teams are no different (Duarte & Synder, 2006). Without trust there is little that can get done. In order to generate trust, we must make an investment of ourselves. Think about it. When you were a child, you spent a lot of time with your best friend. In many cases, it was that investment of time that made the difference. It was not about teambuilding events or even the places you went, but certainly there was an investment of time that helped develop the relationship.

Another important element of this kind of trust is about secrets. A major point about having a best friend is to have someone to tell a secret to. We may not remember all the events we shared with our best friend, but we certainly remember some of the secrets we shared. Furthermore, if our best friend ever told another a secret, that event is etched firmly in our memory. Looking back on these secrets, we might not think they are so important now; after all, does it matter who was a person's crush in seventh grade? Yet at the time, those secrets were more important than anything else in the world.

This sharing about what defines a person is what makes these secrets so important. Sometimes others might guess these secrets, but it is important to keep these ideas at least partially obfuscated to give the secret some deniability if challenged. Secrets become part of what makes people complex. People have deeper feelings, thoughts, ideas, and beliefs that they often keep hidden from others. These concepts are often kept hidden from most others but a person might engage people in the tribe with these secrets.

It has often been said that people have a public face and a private face. In some cases, an individual might even have multiple personas. It is this why the concept of the tribe is important. When an individual engages the tribal relationship rather than just engaging others with their more superficial personas, the individual is less vested in the project. The difficult part is having people interested in engaging at this level instead of at a more superficial level.

Purpose is another important element of the micro-team. Purpose can be temporary or permanent, and the duration should not impact the team dynamic; in some cases the time horizon can seriously impact the perspective of the micro-team. In the example of couples, the time horizon should have a horizon of many years to life. Legally, married couples in society remain together until death or divorce. From a spiritual standpoint, some religions believe that married couples remain together until death. Clearly this kind of time horizon offers the perspective of building a long-term relationship together, working together, having children and raising a family together, saving for a new home or a family vacation. However, in business, few if any micro-teams will remain together on a permanent basis.

When micro-teams are together on a temporary or short-term basis, there is a shift in perspective. The micro-team might have a long-term goal to remain together for the next project, but that is not always possible. From a more realistic approach, the micro-team needs to work in the present and learn how to be successful with the project at hand. Many experts have identified some projects that are hard to close out because individuals see the project completion as the end. In some cases this is true; when a project is closed out the entire project team will dissolve and may not work together again, and this does have an emotional impact upon some. When one individual in a relationship does not want to give up, there is a long protracted make-up/break-up cycle that eventually results in closure; one can find the same in a project.

Equality is another pivotal concept about tribes. Some people might conclude all individuals in the micro-team are considered equals, but this is not always the case. In fact, it might shock most people to realize that this is almost never the case. If total equality were the case for a micro-team, very little would ever get done. What is forgotten is that equality does not always mean that people are represented as equals. Equality is more about each individual in the relationship being considered a participant with equity in the relationship. A tribe will not always have a group of chiefs but there cannot be a group of followers either. If there is a group of chiefs, then there is always conflict and strife as individuals try to assert themselves in the relationship.

A widely circulated story in chain restaurant lore addresses what happens when one brings together a community of chiefs. A certain restaurant opened at which the company had brought together all their best employees from various locations. The idea was that this great new opening could not fail so every possible resource was devoted to this new restaurant. By bringing in the best of the best from all their locations, the organization felt they would create an uber-team that would guarantee success. Everyone was a senior member of the company and all of them had years of experience in each of their areas. The restaurant opening was a complete disaster because all these senior people felt they were the experts in their respective areas and that their opinion was the most important. By bringing together an army of chiefs without an understanding of the hierarchy and goals, the new venture was doomed to disaster.

HOW SMALL TEAMS SUPPORT COMPLEXITY

Small teams can be an important part of complexity in teams. Consider the small team of two hydrogen atoms and one oxygen atom. Separately, these are two separate elements with totally different properties. Oxygen is one of the building blocks of life on this planet, while hydrogen is a particularly explosive element that is abundant in the universe. Yet together these elements create water. This is an excellent example of a small team that comes together to create something more. Apart, these elements have their own individual properties, but together they create something that is another building block of life on this planet.

This analogy might be interesting for a new project manager; however, a project needs more than just a clever point to apply complexity. This example helps us understand that great things can come from simple elements. The concept of the wheel exists in the natural world. Rounder items naturally roll downhill with less effort than nonround items. It is not such a stretch to see that harnessing round items for work would vastly improve a creature able to grasp the concept. Yet round alone is not sufficient for success. To truly harness the wheel, the concept of the axle must also be employed. There is no living example of the axle in the natural world. No animal uses the concept of a wheel and axle for transportation—except for humans. This was a technological leap beyond the natural world. Again, the project manager may wonder how this applies to a project.

Small teams support complexity through accelerated learning and achievement and their skilled use of the useful bystanders. Small teams apply to complexity because it is these teams that actually accelerate learning and achievement in organizations. Often these small partnerships help the team achieve more by offering some helpful competition within an organization. Small and focused teams will often create a network of individuals who are connected to the team in a way that can be helpful later. These would be more than just stakeholders because many times a small team can harness useful bystanders in order to assist with the success of the team.

As in the axle example, it would be easy to conceive that this development occurred within a small team. Although creating wheel and axle may have been a one-person job, once the item was created the individual would have to show it to someone. Most likely the creation would have been regarded with some contempt and no doubt some kibitzing would occur. What is important is that new inventions, developments, and organizational changes do not occur in a vacuum. They always must be done with others. If a team can discuss possible solutions internally without any concerns or repercussions, it can make larger improvements. A small team that can review internal improvements has a better chance of making changes because there is always some reluctance to offer up a new idea to one's boss. If one has a safer place to discuss these kinds of ideas, then these ideas can grow into more than what they started out as.

A project manager can consider forcing small teams into a group by asking team members to partner up with others and keep them advised of the one another's project tasks. This allows individuals to talk to others about progress and how they achieved their task. Not all partners will create greater progress, but it will allow people to interact more with others. This will help keep the project on track as it makes people more accountable and offers a safe haven for new ideas. It will also give the project manager a number two person to go to for information and progress because it will help make some redundancy in the project. Because complex projects can change quickly, there is a possibility that a person might be moved off the project and it would then give the project more flexibility in case of these changes (Cooke-Davies, Cicmil, Crawford, & Richardson, 2007).

Present the concept to the team as each team member needs an understudy for a complex project and complexity. Just like a major theatrical event, every cast member needs an understudy in order to be sure that the production is completed as expected. Consider how an actor rehearses a part; this is a complexity-based system. Individual rehearsal is done however the actor feels best, while production rehearsals are done with others. Many actors go through complicated rituals in order to get into character and to memorize their lines. This process is unique to the individual, although the final result is the same— actors must be prepared to perform their roles. This process is the same for the understudy, who must be as prepared as the main actor, since one never knows when the understudy will need to take on those additional responsibilities.

This is no different for the complex project. Certain safeguards must be taken in order to keep the project on track. Even though some of the participants and stakeholders might change, the path of the project can remain constant (Weaver, 2007). Redundancy in a system is often the difference between success and failure. If a system is subject to a single point failure, then the system is always inherently weak since a failure in any single node can result in a complete failure of the system. Building some redundancy into the system keeps the system from failing in this manner. This does not mean that there are not other opportunities for failure, or that two people will also fail, but it does help make the system more stable. The project manager must consider this carefully as it may cause some concerns in the team but it can also have some significant benefits in the long term.

LEVERAGING MICRO-TEAMS WITHIN ORGANIZATIONS

As discussed earlier, micro-teams are a small group of people (usually two or three) brought together for a specific purpose who communicate among themselves and have some type of leadership structure (Godin, 2008). Sometimes mundane micro-teams focus upon a simple goal and other times it is something grander. Tribes, on the other hand, are a group of people brought together for a similar purpose, are focused upon a goal. The tribe must communicate internally and externally and they must be led.

It is important to have individual goals. In project management it means that the tribe (including all of the micro-teams) must be focused upon the project. So how does one make a project team focus on a project? To understand how tribes work, one can look at a more mundane organization to get some good ideas on how to make it work. We can examine own direct family structure for ideas on how it would work.

To start, it is clear that not all families lead the perfect Brady Bunch existence, nor should we look at that as the norm. One needs to focus on how their family works and does not work. Learning from imperfection is one of the project manager's best resources and should be part of any toolbox. To begin with, a family is a small tribe of people all focused upon survival. Sometimes it seems that every family is more focused on birthday or holiday gifts, and if the family has more video games than the neighbors, but at the end of the day, the family is focused upon survival. When all the material goods are stripped away and one focuses upon what is essential, one will find that it is all about survival.

Consider the many ways your family survives, communicates, and most of all how it is successful and not successful. If we can glean what works and does not work for the project manager, this information can be superimposed upon the group. For example, many families like to write newsletters about their family's achievements and changes to keep in touch with other friends and family. This seems like a rather Martha Stewart kind of metaphor, but it is something that helps keep a family together. Since it is pretty unlikely that one would ever send a family newsletter about only the awful things in life, a family newsletter actually offers the family opportunity to reflect upon both the good and the not so good. During the writing of such an update, one will be likely to put some spin control or life's lessons reflection upon the negative and to highlight the good.

One such family newsletter that was several pages long contained a half-sentence comment about a family member being unemployed for 14 months, but there was a full paragraph about that same family member focusing their extra time on building a boat for the family to enjoy. Now, it may be unlikely that you possess the skills and ability to throw together a family boat, but it brings up the point that when one looks beyond what is apparent, there is often more there. For a tribe this kind of information is essential as it offers a good look into what works for the micro-team. This not only keeps a family together but helps them survive. Many times for the tribe, the adventure is not in the destination, but in reaching the goal.

When the average tribe of four gets into the car for a family vacation with the goal of traveling to the nearest family amusement park, the destination will certainly be much more exciting than the journey there.

The transit is only a means of achieving the ultimate goal and hence the ultimate goal is the sum of all dreams. This kind of situation often causes great friction within a family. Anyone who has been trapped in a car with kids droning *Are we there yet* can understand why some family members have trouble communicating. However, from a tribe standpoint, the project manager must understand whether the destination is the source of all achievement or the road there is the source of achievement. This becomes part of the challenge for the project manager, because the reality is that both the road and the destination can become a source of achievement and the more that the project manager can help the tribe see that, the better focused they will be upon the goal.

There are two great lessons from the example of the unemployed family member who was building the boat. First, the family members' spouse, in retrospect, is about the smartest person on earth. By including that information in the family newsletter; it achieved two important things in the survival of the tribe. First, it reframed the spouse from an unemployed deadbeat who no doubt was asked by everyone and their sister about not finding a job to a family-centered father figure. By the way, asking a person who is unemployed why they do not have a job is not particularly productive and often makes for poor dinner conversation.

Second, by spreading the news that the boat was being built for the family, it forced the person's hand to complete the boat. One can only imagine the responses to this person after the boat project was *bragged* about. The dinner inquisition goes from the twenty questions about interviews and job leads to a friendlier and more casual conversation about the progress of the boat and, more importantly, when can it be borrowed to go fishing. (For the record, the family member is now employed and the boat is finished.)

This becomes a case where a family project (building a boat) has gone from a destination-only project to a road and destination project. This example is actually based more in projects than one may realize. A project newsletter is often a great way to communicate to everyone about the project. It offers a way to let stakeholders and team members, as well as the community, know what is happening with the project. A project newsletter circulated on a regular basis offers a great support mechanism to describe the project to individuals so that they can understand the importance as well as to understand the team involved. Furthermore, the actual tribes involved with the project should be contributors or should create their own circular about the project. This allows for the individual tribes or families to communicate their milestones and their challenges, and to highlight their creativity at overcoming obstacles. It is a natural matter to want to try to paint an appealing portrait of the elements of a project rather than just to have a static report of doldrums detail.

To make a newsletter successful, the project manager needs to make sure that the information is passed along even to those outside the stakeholder community, since one never knows how this might also assist with others to defect to the tribe. A successful community is one that people want to embrace rather than avoid. In addition, it offers a good way to communicate the positive about a project (such as building a boat) rather than dwell upon too many of the negatives about a project (such as the potential of unemployment at the completion of a project). The project newsletter should serve to become the voice of the project and of the tribe in order to make others want to share in the experience.

Beyond the newsletter example, the project manager needs to better understand the tribe and to make the project one in which the road and the destination are both worthy of advertisement. When the road becomes fun, everyone wants to help. This is seen in exciting projects everywhere. If things are going well, everyone wants to be a part of it. So the best thing a project manager can do is make the project envied by others. Allowing others to see what the tribe is accomplishing enables the micro-team to better leverage the organization.

Hence, one of the goals of the project manager is to make the micro-team a much-desired structure within an organization. The project manager also needs to advertise the project to make sure that others see what is being done along the way to the destination. If one can leverage that feeling and excitement of taking an adventure to achieve the goal, it makes the project more interesting and one can be certain that others will help the project become successful as more people find a reason to get involved and to join the fun.

HOW TO BUILD TRIBES WITHIN AN ORGANIZATION

Organizational Culture Attributes

In order to build tribes within an organization, one must examine the different elements that make up a culture. A tribe may be a small group of like-minded individuals focused on a common goal; a tribe shares all the elements of a larger culture (Godin, 2008). In fact, the cultural attributes are greatly enhanced because with such a small group, a single individual's actions can influence the tribe to a greater degree.

Six important elements make up a culture: innovation and risk taking, attention to detail, outcome orientation, people orientation, team orientation, and assertiveness. Leveraging these six attributes can assist in building tribes within an organization, so it is important to understand how each fits with complexity.

Innovation and Risk Taking

Innovation and risk taking is definitely important to the culture of a tribe. The more the tribe feels empowered to take risks and to develop new innovations, the greater the contributions the tribe will make to the team because a team that is willing to improve is a team that others will want to join. Essentially, it is very important that all team members are innovative in their job in order to enhance the project. The tribe must embrace the idea that trying new things is good for the project and must do away with the "I'm right, you're wrong" mentality. People are to be culturally encouraged to find solutions to problems, even when there are not immediate project issues. Many times a team with experience will know the critical points of failure in a project and even more often team members will know the solution, but they may feel helpless to change the outcome. This feeling of pending doom can be corrected by creating a culture that supports innovation. Additionally, if the tribe feels they are supporting one another, they will feel strong enough to challenge any authority that may stand in the way. The tribe must learn to support individuals in a manner that they feel valued, safe, and trusted. So even if their recommendations are disregarded, the tribe will continue to offer more solutions. Hence, even failure must be celebrated as a way of preparing for future success.

Attention to Detail

A tribe must support the details of the project. The highest quality is always important and even minor details must be handled as if they were a crisis. The tribe must support everyone's participation in the concept that every detail is very important (Godin, 2008). If quality is not visible in the project, then the tribe is not contributing anything more than what any other team could achieve. The details are where the problems reside, so if the tribe is focused on looking critically at the details, the team can better resolve small problems before they become large problems. Allowing the team to resolve these problems is at the root of complexity. Complex projects require solutions.

The more complicated the system, the greater the opportunity that a complicated problem can develop as the project grows (Hass, 2009). If the smaller problems are resolved dynamically along the way, then less complicated solutions can resolve the problems that develop. Tribes that are close to the project have an ability to reduce problems in a way that can achieve solutions in a manner that more encumbered individuals might not be able to achieve.

Leveraging this ability is important as it is a way that complexity is at work. Understanding the problem is often 90 percent of the solution, so the more people are engaged, the more equipped they are to be part of the solution. Too often, individuals who are removed from the problem see the solution as adding an additional layer of bureaucracy in order to monitor the broken system. This then allows an individual to intervene to keep the process moving rather than to create a system that functions effectively.

Outcome Orientation

Project managers in general are driven by outcomes. The ultimate goal of the tribe should always be the successful launching of the project. Given that all areas are measured by outcome, the tribe must conform to this reality. Producing results is the single most important factor in a project. Companies demand results, and more importantly, organizations are anticipating results beyond the project's expectation. Careers are made or broken by results. Those who perform under difficult circumstances are rewarded, while those who do not perform at that level are deployed to another location, or worse. If an individual tribe finds that their style, their abilities, their ideas, or their culture does not allow them to excel, they should consider adopting a different culture or moving on to another project that better fits what they can offer.

Too often, tribes feel trapped because there are outcomes that are beyond their ability. The tribe might feel that they are in a fight-or-flight situation, so they may start to complain about all the aspects of the project or they may start to leave the project. Either one is a problem for the project manager because when important resources are sapped from the project, the project will suffer.

People Orientation

Organizations must have a clear customer focus. Tribes must also have a clear customer focus, not only for the external customer but for the internal customer as well. A tribe must do all it can to avoid a caste culture of haves and have-nots. The tribe must do good job of treating one another with the same care and respect with which they expect others to treat them. This is one area where the project manager can make all the difference in the world. If the project manager makes people a priority, the tribe will have no problem doing the same. A project manager can utilize recognition programs, expanded training programs, and better communication; however, a large organization can still easily overlook a tribe. Interestingly, research has shown that the project manager does not have to be perfect at making people a priority to make this a part of the culture. The project manager just has to do all he or she can and to show the team that they are trying to make it part of the culture. This is another element of complexity in which no one thing makes the difference, but it is important to show people that *something* is being done as an important aspect of the culture. The tribe will follow what the project manager models, not what the organization does as a whole, which is an interesting point to keep in mind.

Team Orientation

Teams should be a cornerstone of the culture of a tribe. Team members must trust one another and find that the tribe is important (Godin, 2008). A tribe might not be a unit recognized by the organization, but the tribe must be recognized by others as a functioning unit. It should feel as if speaking to one is as speaking to all, and that where one goes the others should be near. A tribe can be successful even where the team concept does not extend to the whole organization.

There is also a negative side to this since small groups will sometimes create an agenda that serves only them. This becomes particularly destructive and often leads to turf wars. Many groups like to come to agreements that achieve consensus, but turf-sensitive groups use this as a way to mark territory. Complexity can be dangerous in this regard because if it is not contained it can lead to problems. A project manager must do all he or she can to keep the tribe focused on the project's success and not on some alternative agenda.

Assertiveness

Assertiveness is important to the tribe. The tribe's voice must be heard and the tribe must be a stabilizing force for the group. Assertiveness should not be confused with aggressiveness where a person might be considered offensive or not a "team player." It is often hard for the tribe to understand what is an acceptable level of assertiveness. Executive leaders are expected to have this trait, and the organization encourages and rewards leaders with that talent. Yet when others display that trait they are often considered rude or mean and are not always welcomed to participate. Some individuals who display this trait may be shunned but they are often effective and produce quick results. While others are worried about "stepping on toes," the assertive tribe can move swiftly and effectively. To be innovative, one must be quick, and that often means "stepping on toes" and not keeping everyone in the loop.

Tribe members must learn to be assertive because if they do not, their achievements will not be recognized, their actions will not be noticed, and the team will ultimately fade away. A great team that is not noticed will not achieve the standings of a great team in an organization, or worse, others will take credit for the tribe's achievements. A tribe must learn to balance being assertive with advertising its achievements. The more the tribes do, the more that they should let others know—a tribe must learn to communicate the good and the bad so that others know that something is being accomplished.

Building the Tribe

All six aspects are not necessary to create a tribe, but leveraging these six elements will help the tribe not only to exist but to thrive. Furthermore, it will allow the tribe to become a potent force, able to leverage important resources within the organization as well as to leverage complexity. If the project manager can steer the new tribes towards these elements then the organization can become stronger as more tribes are present and move the project towards greater success (Duarte & Snyder, 2006).

Building tribes takes time and resources. Yet these are not always just the project manager's time and resources. Once a culture of a tribe has started, it will take on a life of its own, becoming a complexity-driven organization as individuals see the benefits of a tribe. The more a tribe can achieve and the more reward and recognition the tribe can gain, the more others will copy this form of organization in hopes of achieving their own success. Often the best way to communicate a successful team is to reward other teams. That is often the best advertisement (PMI, 2008).

SUCCESSFUL TRIBES

The Standish Group (2009) has stated that the projects of the future that will be most prolific will be those handled by micro teams of four or fewer members, will have a period of four or fewer months, and will be budgeted in the range of $100,000. Since there is such a focus on this area of project management, it is best to consider what makes for a successful micro team or tribe.

As previously defined, a tribe (micro team) is a small group of like-minded individuals focused on a common goal. This is the starting point for reviewing what makes for a successful tribe. There are five important elements that make up a successful tribe:

1. Understood goal
2. Making more from less
3. Creativity (reward)
4. Examples
5. Eliminate dependencies/Create interdependencies

Understood Goal

The tribe must always start with a common goal, which must not simply be known but must be the material that binds the team together (Duarte & Snyder 2006, Godin, 2008). This recalls the adage about being involved versus being committed. When one evaluates the standard breakfast of eggs and bacon, one will find that the chicken was involved in the project while the pig was committed to the project. Since a tribe is so small, everyone must be committed to the project. The micro team must eat, sleep, breathe, and live the project. The team must consider that there is no tomorrow. Since a micro project is so brief, there might not be a project tomorrow. Projects need to get done and the team must be prepared to move to the next project.

Regarding complexity, each team member must be connected to the goal while understanding his or her role(s) in the project (Hass, 2009). A small team means that all participants will have to wear multiple hats, and this means being able to change hats quickly and efficiently. Complex projects need people who can change and adapt rapidly, and in micro projects this is critical. A short deadline and limited budget means that the team must be prepared to adapt the project in order to achieve success.

To review an extreme case of this kind of thinking, we turn to the great polar explorer from the turn of the century, Ernest Shakleton. Shakleton was committed to reaching the North Pole before any other man on earth (Lansing, 2007). He set out with a team of explorers to reach the elusive South Pole. Several other teams had tried and many had died in the attempt but he was undeterred in this quest. The plan was that the team would sail as close as possible in the ice and then travel the rest of the way in groups of sleds that would be pulled by dogs. The tribe was focused on the goal and they journeyed out toward the North Pole.

Shakleton and his team set out onboard the vessel *Endurance*, but the plan did not quite work out as expected. The ship got stuck in the ice and after waiting months to get loose, the strain of being crushed by ice eventually sank the ship. Once Shakleton knew that the ship was going to be lost, he had to move his plan of adventure to one of survival. He was committed to leading his entire tribe safely back to civilization. Since this was the early twentieth century there were no satellites or other long-range communication devices, so Shakleton knew that he was alone with his tribe. He also knew that he was in the most inhospitable area of the planet and that no one would be launching a rescue party to locate him or his team. Furthermore, prior teams that had attempted to reach the pole had ended in almost 100 percent casualties (Lansing, 2007).

Shakleton then systematically and methodically did everything to keep everyone alive and to make sure that everyone was rescued. The hope for survival was to travel as far as possible by sled to a point where one of the lifeboats could be launched on a week-long journey to reach a tiny island that had a whaling outpost (Lansing, 2007). Along the way the team had to continuously make tough survival decisions, and at any time a bad decision could have cost the entire team or at some team members their lives.

Making More from Less

Tribes excel at making more from less. In other words, the micro team must learn to wring every resource available in order to maximize their value. The tribe must learn to scrounge. Sometimes it means making unconventional decisions in order to achieve the desired results (Duarte & Snyder, 2006).

This is another case where one can use the example of Ernest Shakleton and his polar expedition. After the *Endurance* had sunk, the entire team was committed to doing more with less. With the loss of their ship, they lost their last connection to civilization. In order to survive and to have a small group reach civilization to arrange a rescue for the rest, the team needed to transport the lifeboats across the ice, carry all the food that was possible, bring other sailing equipment from the voyage to the whaling outpost, and manage this while battling the elements and extreme weather. A team value was always to travel as light as possible, but given the circumstances the team had to balance what was necessary with what would be left behind. Items were forced into multiple uses as the team had to work creatively with the tools, equipment, and rations they had. One can only imagine the sacrifices that such a team had to make in order to achieve success.

In the end, the entire team was committed to the new goal of survival. After months of trying and continually struggling with the adversity of starvation, the weather, and the continuous freezing cold of living on ice; the entire team was successful in having the smaller group reach the whaling outpost. Once Shakleton arrived, he was able to gather support for a rescue. However, the team that remained on the ice was in such a difficult place to reach that they were not finally rescued until months afterwardand three attempts. The first two failed due to pack ice nearly trapping the rescue vessel. In the end, every one of the team was successfully rescued.

Creativity (Reward)

Business is changing quickly and because of this change, individuals are forced to adapt just as rapidly. Consider that, on average, businesses have to replace half of their employees in four years, half of their mid-level managers in five years, and half of their senior executives in seven years. Organizational roles are changing so rapidly that individuals should expect that their career, trade, or role would have a lifetime of ten years or less. Individuals who fail to adjust their roles will find their skills outdated in several years.

Currently, organizations will train a person for a particular role; however, there is no guarantee that the role will have significant application outside of that organization. These acquired skills might not even be valued after a few years. Although a person may become skilled in a field, there is no guarantee that the field or the organization will outlive the employee. Of the top 100 companies in the U.S. fifty years ago, very few are in existence today.

Given these changes in business, a tribe must learn to be creative in order to be successful (Rigby, 2009). Since business is changing so fast and understanding that yesterday's skills might be no match for tomorrow's challenges, a micro team must learn to be creative in order to adapt old skills to a new environment. These micro projects will have a short timeline and skills and technology might not change during the time of the project, but there is actually no guarantee that the goal will remain static. It becomes necessary for these teams to be creative in making sure that they accomplish their goal in time.

Consider that a software development company starts to develop a new software program in the current existing operating systems and that company moves forward with the latest tools and techniques. The project has a timeline of six months and the new program has been announced to the media and the expectation is that the new offering will be successful. Halfway through the project, there is a new patch for a couple of operating systems that was done in order to correct some inherent security issues within the operating system. Additionally, new development tools are now available that can drastically change the efficiency of program executables. These new developments concern the marketing department of the software house since by the time their new program is released it might already be considered old technology and might not be compatible with the updated operating system. The question becomes: what should the team do?

This is where creativity needs to come into the micro team. The small team should be flexible enough that it can address these new challenges, and the organization should be savvy enough to offer to reward them for adjusting the project to meet the new needs of the market. This is where the project leader needs to balance the needs of the team and the needs of the organization in a creative way to achieve the desired results.

In this case the desired results are to offer a new program that uses the new tools and technology that is also compatible with the latest operating system patches. The reality is that the team will have to take on additional work with a short deadline in order to meet the already agreed upon launch date of the program. This means altering the development plan without adding additional time.

Examples

A tribe needs to have both negative and positive examples. Sometimes negative examples must be made of people who fall off the path. It does not mean having to fire a team member, but it could end up being just that. A successful tribe needs to understand that there are penalties and consequences for incorrect or improper actions. A tribe also needs positive examples, where an individual is recognized for superior effort.

Although it is important for those who are viewed as leaders to shape the culture by reinforcing positive examples, it is also important that all the members of the tribe do the same. The group is too small not to have everyone recognize positive effort. The more everyone recognizes this kind of behavior, the better the entire group will be. Successful organizations are those in which everyone recognizes the efforts of individuals, rather than just a few. If everyone sees what others are doing, then there is greater motivation to do more. Consider this as peer pressure for over-achievers. The more people observe that recognition is bestowed upon those who put forward superior effort, the more the tribe will be pushed to achieve more.

The inverse of the situation is equally important as those who do not carry their weight and do their fair share are not as valued and those individuals need to be recognized as well. This is a more delicate situation as the negative example should be addressed in private, but there must also be a public declaration to the rest of the tribe. If a low achiever is only addressed in private, the rest of the tribe will feel that nothing is being done. This makes for a difficult situation but it needs to also be addressed.

Many managers feel that this could cause problems for the organization if a negative example is made. Yet, there are ways to address this publicly without malice or prejudice. An example of this would be by tracking milestone achievements for the tribe. If one publicly communicates the efforts of each individual then there is no malice, it is just representing the facts to the team. Those that are negative examples will appear just as prominently as the positive examples and this will force people to reflect upon the situation. The individual that is having difficulty will need to consider their role in the team and either put forth greater effort in the future, or they will have to consider being replaced in the tribe. This kind of reporting will make it obvious to everyone on the team and will put peer pressure on everyone to achieve more.

> Practical Tip: Make your tribe the one that others envy. If a tribe is particularly coveted, not only will it attract the best but it will lure "groupies" who will want to be a part of history. Think of how many people were in the bands that played at Woodstock (made history). Consider how many people went (part of history). Consider the larger number of people who wished they went (missed history). Finally, consider how many people that one event touched (touched by history).

Eliminate Dependencies/Create Interdependencies

The largest dependency of any tribe is upon the leader. If a tribe feels that they need to run every idea past the leader and wait for judgment, then the leader needs to eliminate this dependency. This kind of dependency is counterproductive to achieving complexity within a tribe.

The tribe needs to operate as independently as possible. To achieve this end, each tribe member must feel comfortable enough in their role and in their decision-making process to move as quickly as possible through all the tasks assigned. If individuals understand their roles and have done the necessary research regarding the task, then they are the best equipped to make the appropriate decisions in that aspect of the project. If the individual is not the best choice for the decision, then that tribe member should not have been given that responsibility in the first place.

The tribe must learn to create efficient chains of interdependencies. It should be no surprise to any project manager that the assembly line is one of the most important management principles and is often one of the most widely ignored for white-collar tasks. Most organizations consider the assembly line to be an industrial age technique that only applied to manual labor or repetitive tasks. The tribe must learn that the assembly line is as applicable to project management tasks as it is to manual labor.

The concept of the assembly line is as follows: repetitive work allows for the individual to specialize in a particular ask in order to improve efficiencies across the entire assembly line. So, by examining the traditional assembly line role, the concept is that an individual will become more skilled over time by handling the same task. A traditional example is a person working in a car factory attaching a door to a car frame. That individual would be responsible for attaching doors repetitively in order to keep the assembly line moving. What is interesting is that the organization is comfortable with assigning that individual responsibility beyond their position within the hierarchy. The individual responsible for attaching the doors to cars could be responsible for thousands of dollars of inventory in their area (they are responsible for a lot of car doors), but they would probably need three signatures to order a new pair of work gloves.

The tribe must learn to do the same thing. The person who is most capable should be assigned a repetitive task without regard to the hierarchy. For example, in a linear project team, the project leader would be expected to make the formal report out to various stakeholders. In a complexity-driven tribe, the person who should make the report out is the member of the tribe who would be most successful at this task. A tribe may have a person connected to other stakeholders who has better training at public speaking or who just does a better job at a task like this. This does not mean that the project manager has abdicated responsibility; it just means that the spokesperson for the tribe need not always be the hierarchical team leader. This repetitive task could be better achieved by someone else in the tribe, so that person should be charged with that task. This will make for a more effective team because it will have the best person handling each task. Just like an assembly line in a factory, this will lead to greater efficiencies. One needs to consider this at every opportunity in order to create a more effective organization (Rigby, 2009). By focusing on a person's strengths, it can make for a complexity-driven tribe that is more successful than having a linear-based team.

> Practical Tip: Focus on strengths. Complexity thrives upon a person's strengths because it coaxes more from the existing individual talent. Complexity is what allows the mind to create a symphony from an assembly of existing notes.

Thus, tribes need to consider all of these points in order to be more successful. A tribe should not be limited to linear thinking because often this keeps the tribe from leveraging the internal strengths. What advantage is it to have skilled individuals if they are forced to funnel their knowledge through people with less experience? Great people can achieve great results if they are allowed to use their skills.

CHAPTER SUMMARY

Tribes have the potential to be units of greater achievement if they are not shackled by hierarchy. Small teams that focus on individual strengths and leverage those opportunities will be those that achieve greater success in the future. Many organizations talk teams and then try to jam experts into a formal hierarchy where the individuals with complex skills are diluted by management. The strength of tribes comes from the individuals who have ability greater than others on the team. A balanced tribe is one that can maintain a small group of experts for a greater purpose.

A common and very notable example of this is a rock band. A band consists of a team of specialists who come together to make music that is pleasing to groups of people. Although each band member has his or her area of expertise, this does not guarantee success. In fact, most bands struggle to stay together and most never reach a level of success where they are recognized on the world stage. Yet there seems to be a continual influx of new artists that gain fame and notoriety. Some of these tribes are famous for decades and some even continue to be famous, even after their death. One must consider this when building a tribe because some of the greatest tribes are those whose fame even grows even after the team is long gone.

> Practical Tip: Make your tribe the stuff of legends. Anything less is not reaching high enough.

REFERENCES

Cooke-Davies, T., Cicmil, S., Crawford, L., & Richardson, K. (2007). We're not in Kansas anymore, Toto: Mapping the strange landscape of complexity theory, and its relationship to project management. *Project Management Journal*, *38*(2), 50–61.

Duarte, D., & Snyder, N. (2006). *Mastering virtual teams: Strategies, tools, and techniques that succeed* (3rd ed.). San Francisco: Jossey-Bass.

Godin, S. (2008). *Tribes*. NY, NY: Penguin Group.

Hass, K. (2009). *Managing complex projects: A new model*. Vienna, VA: Management Concepts.

Lansing, A. (2007, 1959). *Endurance: Shackleton's incredible voyage*. NY, NY: Basic Books.

Merriman, K., Schmidt, S., Ross, G., & Dunlap-Hinkler, D. (2004). Virtual intraorganizational authority relationships: Implications for trust, support, and influence. *Academy of Management Proceedings*, 5.

Project Management Institute (Ed.). (2008). *A Guide to the Project Management Body of Knowledge—Fourth Edition*. Newtown Square, PA: PMI.

Rigby, D. (2009). *Winning in turbulence*. Boston: Harvard Business Press.

Standish Chaos Report. (2009). Standish Group.

U.S. Army and Marine Corps (Ed) (2007). Counterinsurgency Field Manual. Chicago: University of Chicago Press.

Weaver, P. (2007). A simple view of complexity in project management. 2007 PMOZ Conference Keynote address.

Chapter 18

Dealing Appropriately with Change

Figure 18.1 Everyone is uncomfortable with change, so it is always a challenge to become more comfortable with the unknown. Watching birds in flight on the other side of a window offers perspective on change. We are most comfortable with change when it is observed from a distance and when we are not subject to the environment. We feel safest when we can watch change from the security of our own personal fish bowl.

Project managers deal with two types of change during a project: organizational change management and potential scope change, sometimes referred to as scope creep. Organizational change management is difficult to implement in an orderly and sequential fashion. Maybe it is not meant to be done in such a fashion; especially when an enterprise-wide project or program is being implemented. Maybe it is time for parts of the organization to self-integrate and self-evolve. This will be evaluated through the lens of complexity.

Scope change has to be managed very carefully within a project. Hence this area of the project should not be allowed to be influenced by chaos. The Spiral development process, in the world of software, fringes on complexity. Each spiral starts the development of the requirements but new requirements may soon blossom. So, are these new requirements or are these items needed to meet the original specifications?

CHANGE MANAGEMENT

The one value that is difficult to maintain is the value of change. Most organizations do not value change, and some organizations fear change. Unfortunately, change is the one aspect of business that will happen with certainty. Sometimes change is planned and other times change is unexpected. An example of this resistance can be seen by the reaction that employees have toward outsourcing. "Outsourcing is frequently accompanied by employee resistance. For most hourly employees, and many managers, outsourcing is synonymous with job loss or change" (Useem & Harder, 2000, pp. 25–36). Most individuals do not see change as an entrepreneurial opportunity, most organizations view change as an "evolution and revolution" where "management sprawl leads to a series of confrontations among the layers of management" (Brown, 2000, pp. 52–55).

The project manager should adopt a transformational leadership style for organizational change projects. The transforming leader commits people to specific actions, creates leaders out of followers, and can convert leaders into active agents of change (Bennis & Nanus, 1997). For a transformational leader to be effective in times of change, that person must have his or her own leadership voice, must own the change, and must have his or her own vision or philosophy to help others through the change.

The project manager who communicates a vision and values offers followers a road map that will give guidance regarding the decision-making process. Since leaders will not always be present, followers must be prepared to make critical decisions based upon the vision and values of the leader. "The best leaders are perceived as having highly developed interpersonal skills. They are warm, open, and forthright, and their attitude toward employees is characterized by words like 'empowering,' 'supportive,' and 'benevolently paternalistic'" (Hemsath, 1998, pp. 50–51). Interpersonal skills become the voice of the leader. These interpersonal skills allow the leader to issue edicts and orders, but they are also the vehicle to offer advice and to support the decisions of others.

Leaders must master "the skill of communicating clearly—being able to communicate in a way people can easily understand and accept" (Weiss, 1999, pp. 6–9). Unlike business culture, which is a learned skill, a company business philosophy must be modeled to those within the organization. A common way that leaders perform this goal is through public proclamation.

Business philosophy is based upon the leader's vision to go beyond the clever spreadsheet and market forecasts to develop new beliefs that motivate employees, customers, and stakeholders. "Leaders are the most results-oriented individuals in the world, and results get attention" (Bennis & Nanus, 1997, p. 26). A leader must learn to value the individual while instilling a successful philosophy upon the business venture. According to Hillman (1996), "As long as we regard people in terms of earning power or specific expertise, we do not see their character. Our lens has been ground to one average prescription that is the best suited for spotting freaks." This means that many leaders are skilled at identifying those who stand out, those who are well above average, and those who are well below average. This was important to organizations in the past but is not the role of the transformational leader. The transformational leader is not looking to identify the troublemakers, but a leader of change moves everyone in the same direction.

Understanding the changing cultural forces that influence customers, employees, and stakeholders is critically important in business growth. The companies that understand the evolving expectations of employees, customers, and stakeholders will flourish. Companies that can meet these requirements will be the ones that become successful in the future, while those that fail to do so will shrivel (Duarte & Snyder, 2006). Effective leaders have a business philosophy that is clearly understood and communicated. The philosophy must be composed of articulated values that support the goals of the business organization. "It is assumed here that if core values are neglected in the policy deployment process, it will never be possible to achieve business excellence" (Dahlgaard, Dahlgaard, & Edgeman, 1998, pp. S51–S55).

Leaders should strive to stand out in a crowd by understanding the business culture of the organization. Leaders should instill a philosophy that supports their business goals and objectives. As an example of this philosophy of self-expression, Hillman offers the acorn theory, which states, "every person bears a uniqueness that asks to be lived and that is already present before it can be lived" (Hillman, 1996, p. 6). Leaders must strive to discover their inner vision or inner voice, and then have the courage to express these thoughts within their business organization. Hillman contests that this inner vision is the discovery of our daimon, which we have been given before we are born (Hillman, 1996, p. 8). Bennis claims that this inner voice is what gives internal direction to great leaders (Bennis & Goldsmith, 1997, p. 24). Direction comes from those who stand apart from the crowd, which means to embody values that may not fit today's fashion. Leaders who stand apart from others offer a different view of the future, a view that others might find appealing.

Furthermore, transformational leaders must learn to modify their personality traits that conflict with the company philosophy (Weiss, 1999, pp. 6–9). This statement indicates that being a good follower is as important as being a good leader. Leaders must achieve the paradox of being homogeneous and unique at the same time. The difficulty is always in formulating uniqueness while still following the company culture. This runs contrary to the opinion that leaders are trailblazing individuals with a guiding vision. This also runs contrary to many a company culture where those who survive change the best are those who remain sheltered under the middle bulge (the average) in the bell curve.

Project managers should learn to make a difference by instilling ownership and camaraderie. "Leadership that emphasizes negotiation and coordination over authoritarian strategies to facilitate the internal and external communication" (Kent-Drury, 2000, pp. 90–98) will create the successful organizations of the future. Leaders must successfully listen to the voice of the organization, and then to echo these ideas in their own organizational voice. Giving voice to others is important to any organization that is willing to learn and change (Heifetz & Laurie, 1998).

Why is transformational leadership important to complexity and organizational change? The transformational leader understands himself or herself. This leader does not need to feed an ego but is eager to see a transformation and growth of the followers and the organization. The transformation does not have to be done by the leader and this is where complexity would play a role.

The project manager on a large project or program should set the goals and tone for the organizational change as well as some overarching milestones. Through the project manager's leadership, mentorship, and educational opportunities provided to the team, the project manager has provided tools for each of the "open systems" to enact the organizational change, as needed.

Some of those open systems may go into total failure (just like a disturbed ant colony or beehive), while others successfully implement the change. If the failed system is allowed to reorganize with the proper leadership—not the project manager or other senior leadership—it may actually be more successful than the organization that initially implemented successfully. The employees who pulled together to overcome the failure now have a vested interest and a success story to share. Some may see the failure as a weakness on the part of the project manager, but it is actually a learning experience for many and most likely did not have a major effect on the project's cost and schedule.

Organizational changes have different impacts on individuals as well as on organizations. The impact may be driven by what the project is implementing, how fast the project has to implement it, how many employees will be displaced or fear they will be displaced, and other factors outside the control of the project manager. Each of these factors needs to be assessed by the project manager and the team implementing the organizational change. At times, it is much better to let the chaos handle the change.

Understanding transformational leadership is also necessary for change management in the avoidance of scope creep. Complexity theory in this area is limited. The areas where the edges of complexity theory might be effective would be in the areas of spiral software development or other rapid software prototyping.

Some would argue that spiral software development or any type or rapid prototyping is high risk and therefore should not be allowed to be in chaos at any time. Spiral development can be high risk and should be controlled for configuration of the code. The user and the developers need to have the freedom to constantly review and tweak the spirals to ensure the needs of the end users are being met. That implies that the initial spirals will be chaotic and difficult to envision. The brainstorming sessions may be long and contentions. The change control boards of the project should only stop those features and new requirements that are truly out of scope. The project manager should only step into the development if the schedule is running behind by a predetermined amount. This should be an amount established by the chief architect and the project manager and approved by the sponsor.

The project manager must be comfortable with indecision and with not knowing what is going on except at a macro level. Complexity theory drives this for large projects. This will be especially true on large software projects where rapid prototyping is used.

HOW COMPLEXITY CAN IMPROVE CHANGE MANAGEMENT

Complexity should enhance the focus and leadership of the project manager. The project manager needs to enhance his or her transformational leadership skills. This enhances the abilities of the project manager to focus the project's goals and vision for organizational change management.

These goals and vision are then communicated to the various organizations affected by the project's change. Some of the organizations or open systems will be on the fringes. These organizations may have been deliberately left out of the project's charter because they were not considered to be affected but in reality they were; the organization was just too small to be considered, or the organization was overlooked.

Each of these organizations will react differently when it is affected by the change. For some it may be a cataclysmic change, such as someone kicking the anthill several times. In other organizations it might be equivalent to several scout ants being stepped on in the line searching for food. To the ants this is not a major event. The ants regroup, make sure the intruder or harm is gone, and then go about their business. The ants may have to redo the line or carry off some dead ants, but they are back in business soon.

For the organization that experienced a cataclysmic change, it may not be as easy as rewriting a few processes and going back to work. This team may have to reach out to other organizations. This would be the same as the different types of ants within the hive reaching out to one another, understanding what needs to be done. The employees in this affected team have all experienced the same event. As such they are the best equipped to fix the occurrence and reach out to the right individuals or leaders for help. This team should be the decision maker of its own fate (Cooke-Davies, Cicmil, Crawford, & Richardson, 2007).

The project manager should determine when and if his or her intervention is necessary in matters of complexity. Project managers are individuals who want to solve problems and have a tendency to want to jump in and resolve or control. However, in situations where complexity theory is the overriding practice, the project manager may make the situation worse. Before deciding to come in and resolve the problem or issue, determine if you are really the correct person. Is there someone better suited, does there really need to be someone interjected at this time, or will the situation correct itself in due time?

INSTANCES OF CHANGE THAT ARE APPROPRIATE FOR COMPLEXITY

Complexity helps in areas of project management that are ambiguous or are on the fringes of the project and may not have great effect on the project. Conversely, there may be areas of ambiguity that are highly complex and will have great effect on the project and should also be left to complexity. The project manager may not even understand or know when some of these instances are happening in the realm of complexity.

However, in the area of organizational change, the project manager should be able to determine what areas to leave to complexity. Organizational change normally is an area that affects the culture of the organization and the lives of individuals (Byrne, 1998). Detailed organizational plans normally cannot be put together at the beginning of the project. They are generally tailored to the communications, resourcing, and training plans. As mentioned, all these plans may be highly volatile, myriad in politics, and changing daily. The project manager would be better served by providing high level goals and visions to the organizations and leaders of the organizational change management. Keeping the organizational change team leaders informed and allowing them to coalesce the chaos would be of better use to the project manager.

The area of organizational change is not an area of black and white that project managers are used to observing and measuring. The area of organizational change is a place of changing people's lives and changing an organization's culture, which normally cannot be done instantly without major ramifications (Hass, 2009). The project manager should ensure the schedule allows enough time and consequently enough cost for this important aspect of a project. Too many in leadership do not pay enough heed to organizational change. Senior leadership theoretically understands that organizationally leadership takes time—studies show three to five years. This is forgotten when the change happens in an organization that they lead. This is where the project manager adds value. The project manager must and should constantly remind leadership that the organizational change will continue after the project is in place.

Projects, especially IT projects, require brainstorming. According to the 2009 Standish report, IT projects continue to fail at an alarming rate. Project managers and their leadership need to take note that traditional methods are not working for these projects. Of note, the project manager's incompetency was not listed in the top ten reasons for failed projects.

Project managers should value different techniques that encourage the end user and the technical team to interact sooner rather than later in those areas of software development where there is much ambiguity. This is normally prevalent in the development stage when adding functionality to the requirements. This requires a lot of brainstorming and interaction between the software programmers and the end users.

Those not familiar with rapid prototyping such as spiral would say that the "bull pen" environment is utter chaos. It actually is. When chaos meets reality it ensures that the functionality and requirements developed are within the scope of the contract (for client facing) or scope of work/charter (for internal). This would be the scope management aspect for the project manager. It is definitely a fine line for the project manager to control and manage.

INSTANCES OF CHANGE THAT ARE NOT APPROPRIATE FOR COMPLEXITY

Complexity is helpful with understanding and working through certain aspects of project management, there are several areas where complexity and possible ambiguity is not necessarily the best approach. Given that a transformational leader is often allowing social systems to operate in a manner that works best and often with little direct interference, this is not a reason to operate in this manner at all times. Complexity describes many systems; however, it is not a silver bullet for all situation. Just like contingency theory and other traditional methods in the *PMBOK® Guide* (PMI, 2008), complexity has distinct applications that must be understood in order to be deployed effectively. There are still times where structure, format and direct cause and effect relationship exist and in those cases complexity is not always the best solution.

Complexity is not a good solution for project areas where definitive structure is necessary. If, for example, a project manager is faced with the organizational challenge of where individuals are unclear about their roles, then complexity is probably not a good strategy to deploy. In times where structure needs to be enforced or re-enforced, complexity offers too much ambiguity. Many individuals and many teams have trouble with working in ambiguous situations because it often leaves them without a clear direction. If a project has difficulty with focus, then complexity should not be applied. Once there is a strong structure and the project leader is trying to get better results and allow more creativity in the project and everyone is comfortable with their roles and with one another, then complexity becomes effective. To this end, complexity should be seen as an advanced technique that can yield good results once there is a strong culture and inherent understanding of individual requirements. If individuals are unsure of their role, the culture, or the requirements of the project, then injecting complexity will likely destabilize the project and cause greater distress.

Complexity is not recommended for the change control board, scope control, or contract changes. Those items that are subject to law, may amend a contract, or have major effects to the cost or schedule should be carefully considered prior to considering for complexity. Even under these circumstances, a highly ambiguous item may be subject to complexity initially and then may be brought under a more traditional approach later.

CHECKLISTS FOR COPING WITH CHANGE

For items traditional to change management, the project manager most likely should keep them within the realm of traditional project management. Normally, training is not associated with ambiguity; therefore, the project manager should not consider training a candidate. However, those items within change, whether scope or organizational, that fall within the realm of ambiguity, the project manager would be best served by contemplating complexity as an alternative.

Complexity does not mean ignore, nor does it mean delegate. The theory understands that there are areas the project manager will not be able to control, understand, or even realize the events will happen. It is different than risk, which was covered in chapter 10. Complexity theory revolves around humans within open-system organizations that are

Priority Change Matrix

	High Value Change (More sales or savings)	**Low Value Change** (Possible sales or savings)
Deadline (High urgency)	1. Do it now!	3. Not a priority
No Deadline (Low urgency)	2. I really should	4. No need to rush

Notes:

Figure 18.2 The Priority Change Matrix Form.

constantly changing, evolving, and interacting (Byrne, 1998). In the area of change management, the project manager needs to provide the guidance, goals, and vision to the project team up and down the chain of command. This ensures that the team and senior management are moving forward toward the same goals. As with complexity, some on the fringes may not receive the word and will not implement or will implement differently or poorly. The project manager may or may not become aware of these implementations, which may or may not affect the project. When and if they do, the project manager should assess them for lessons learned.

Figures 18.2 and 18.3 show two possible forms that will assist in making change an important aspect of the project. The first is the priority change matrix. Sometimes when confronted with multiple required changes, it becomes difficult to determine which change should take priority. This form is a way to review all the available changes in a manner that allows comparison of the relative values of these changes. The matrix is set up for sales or savings as the drivers for the change, but change can be driven by other factors such as culture or process, so one can use a similar matrix with culture or process as the key driver when determining the relative importance.

Once the changes have been priorities, the Change Objective Planning form is a way to establish a process to ensure the change is completed. This form can be used as a way to document important changes in the

Change Objective Planning Form

Employee Information			
Employee Name:			
Employee ID:			
Job Title:		Department:	
Manager:			
Date:		Change Period:	to

Instructions

1. Change Objective. Briefly describe each change objective and when the goal/objective should be met or accomplished.

2. Measurement. How will the change objective be evaluated? (use quantitative measures such as % or dollar increase in revenue or market share and/or use qualitative measures which are descriptive of criteria.)

3. Importance. Rank the change objective as Essential, Important, or Desirable as follows:
Essential - The change is required for job performance
Important - The change is helpful for job performance
Desirable - The change is an asset for job importance

1st Change Objective			
Description:			
Measurement:			
Importance	☐ Essential	☐ Important	☐ Desirable

2nd Change Objective			
Description:			
Measurement:			
Importance	☐ Essential	☐ Important	☐ Desirable

3rd Change Objective			
Description:			
Measurement:			
Importance	☐ Essential	☐ Important	☐ Desirable

Figure 18.3 The Change Objective Planning Form.

organization. Something this formal would be necessary if systems or processes will be changing during the project. The form is a way to codify and direct the change in a manner that is accountable and quantifiable. This will also serve as a good document to support required changes within an organization.

Once the change objective, measurement, and importance are determined, the change date period must be determined. The project manager needs to make sure that the change is completed as required. In most cases, the follow-up will be apparent as the change is either happening or not, but making the change a priority and connecting it to performance will make the change important enough that people will take notice and address it.

CHAPTER SUMMARY

After several years of increasing successes in IT projects, the Standish Chaos Report showed a dramatic decline in IT project success in 2009. The traditional method of project management appears not to be working, although a case could be made that this is just in certain areas. Change management is one of those areas that have a high degree of ambiguity.

The project manager for this area should have a transformational style of leadership. This will enhance the project manager's ability to articulate and to provide the confidence to the team to do what is necessary to implement change management. The project manager is the icon of value add for organizational change. Leadership needs to understand the complexities of organizational change and how the changes will continue after the project is implemented. Those in charge of organizational change and those within the organization will also need the benefit of communication.

The project manager is the vessel of communication for organizational change. Organizational change has occurred for as long as modern projects have been around. Interestingly, the *PMBOK® Guide* (PMI, 2008) does not address this part of the project. This could lead the project management community to the conclusion that this area is highly ambiguous and project managers have not come to consensus to include it in the *PMBOK® Guide* (PMI, 2008), or that it is not of concern to the project manager. The latter hypothesis can be rejected since the 2009 Standish Chaos Report does not list incompetent project managers in the top ten reasons for failure.

REFERENCES

Bennis, W. & Nanus, B. (1997). *Leaders: Strategies for taking charge* (2nd ed.). New York: HarperBusiness.

Bennis, W. & Goldsmith, J. (1997). *Learning to lead* (updated ed.). Reading, MA: Perseus Books Group.

Brown, K. L. (2000, September). Analyzing the role of the project consultant: Cultural change implementation. *Project Management Journal.*

Byrne, D. (1998). *Complexity theory and the social sciences: An introduction*. New York, NY: Routledge.

Cooke-Davies, T., Cicmil, S., Crawford, L., & Richardson, K. (2007). We're not in Kansas anymore, Toto: Mapping the strange landscape of complexity theory, and its relationship to project management. *Project Management Journal*, *38*(2), 50–61.

Dahlgaard, S. P., Dahlgaard, J. J., & Edgeman, R. L. (1998, July). Core values: The precondition for business excellence. *Total Quality Management.*

Duarte, D., & Snyder, N. (2006). *Mastering virtual teams: Strategies, tools, and techniques that succeed* (3rd ed.). San Francisco: Jossey-Bass.

Hass, K. (2009). *Managing complex projects: A new model*. Vienna, VA: Management Concepts.

Heifetz, R. A., & Laurie, D. L. (1998). The work of leadership. In Harvard Business School Press (Ed.), *Harvard business review on leadership* (pp. 171–197). Boston: Harvard Business School Press.

Hemsath, D. (1998, January/February). Finding the word on leadership. *Journal for Quality and Participation.*

Hillman, J. (1996). *The soul's code: In search of character and calling.* New York: Warner Books.

Kent-Drury, R. (2000, February). Bridging boundaries, negotiating differences: The nature of leadership in cross-functional proposal-writing groups. *Technical Communication*.

Project Management Institute (Ed.). (2008). *A Guide to the Project Management Body of Knowledge—Fourth Edition*. Newtown Square, PA: PMI.

Standish Chaos Report. (2009). Standish Group.

Useem, M. & Harder, J. (Winter 2000). Leading laterally in company outsourcing. *Sloan Management Review.*

Weiss, W. H. (January 1999). Leadership. *Supervision.*

Part V

ADVANCED TOOLS FOR MANAGING COMPLEXITY

Part V will provide tools based upon complexity techniques for various aspects of the virtual project. This section will review complexity at various organizational levels. Once the practitioner is comfortable with applying complexity at a certain organizational level, this will help focus the efforts in the appropriate area. These tools include traditional checklists as well as interactive evaluations to determine the current status of an organization and what complexity can be used. Furthermore, there would be a section to assist a practitioner to apply complexity as a solution to certain project management issues. Complexity can be used as a solution to address a project that has encountered issues. Finally, Part V closes with an analysis of the future of complexity theory and its application to project management.

Chapter 19

Complexity Tools for Organizations with Virtual Teams

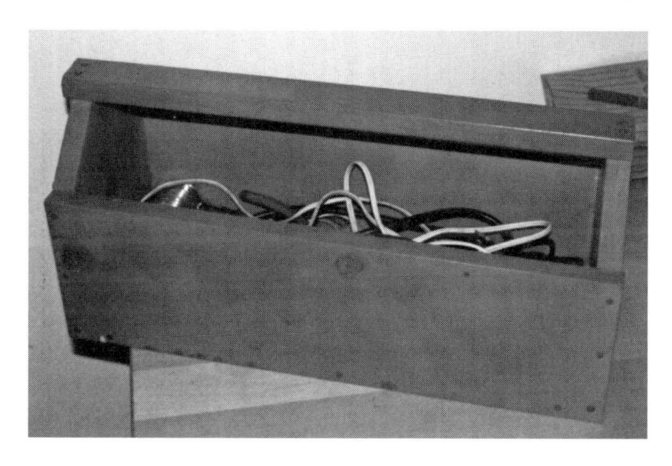

Figure 19.1 The tool box is a metaphor for the tools available to the virtual project manager.

COMPLEXITY LEADERSHIP TOOLS

Virtual Leadership Complexity Process Chart

Leadership is about helping a group of people go from the known to the unknown during challenging times.

Figure 19.2 explains graphically how leadership can be broken down into a series of steps and stages in order to apply complexity within an

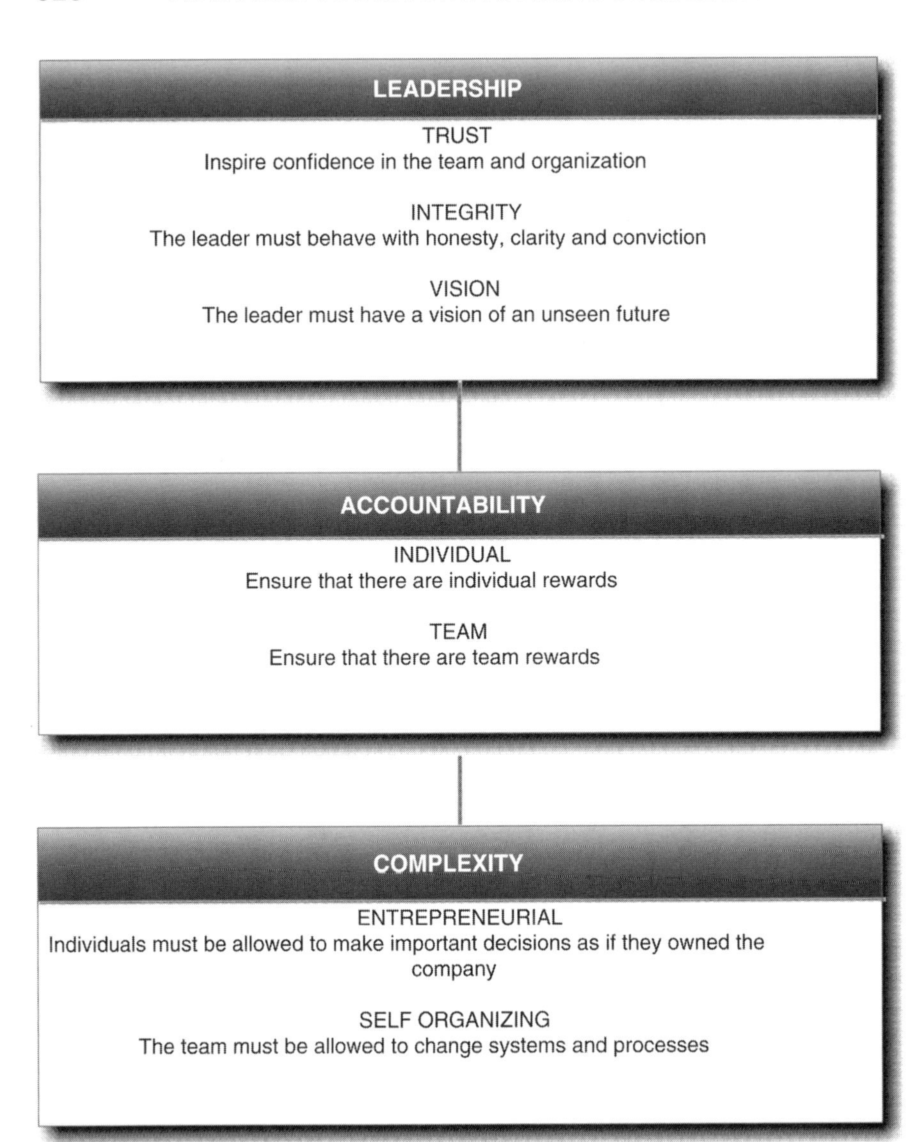

Figure 19.2 Stages of leadership

organization. The important consideration is that the leader must practice letting go of the reigns in order to allow the team to achieve greatness.

Virtual leadership can be broken down into three important elements: trust, integrity, and vision. The virtual leader must learn to inspire belief in a culture and environment so that individuals will be willing to perform new and difficult tasks (Duarte & Snyder, 2006). Often a leader will assist in this process by performing the tasks and then teaching others to do the same. There is value in asking others to do no more than what the leader has already done. It not only is inspirational, but it allows others to see that the leader is willing to undergo the same hardship as others.

Trust

Trust is about creating a reciprocal relationship between the leader and the follower (Lipnack & Stamps, 2000). There must be mutual respect coupled with a mutual understanding. Trust can be described as something that people will recognize when they see it rather than being a specific set of elements.

Integrity

A leader must display honesty and behave accordingly (Bass, 1990). Leaders must be those who others want to follow. Integrity is about others believing the leader and the leader believing that follower. This relationship is important because integrity is defined by the relationship more than just the actions of the leader.

Vision

A leader must have a vision of the future. A leader must be able to see what others cannot imagine and them to that place where even the leader might not have been before (Bennis, 1994). A leadership vision is about building something new out of people and products that are already known.

Accountability

A leader is accountable to the project, the stakeholders, and to the team. A leader must behave in a manner that is culturally acceptable while also achieving what is expected by others (Hamel, 2009).

Individual

A leader must create individual accountability in team members. Team members must learn to keep to their promises and remain aligned with the project plan. Accountability does not always mean that everything will happen according to plan, but it does mean that people will be responsible for their actions or inactions (Hamel, 2009).

Team

A leader must create team accountability where actions of the team have positive and negative ramifications. Team members must be accountable to one another, as everyone must complete their tasks on time in order to make the project a success (Tichy, 2004).

Complexity

Complexity is about the symbiotic relationship of social systems and the nonlinear thinking typical of human systems (Jaafari, 2003). An effective complexity-based micro-team will be flexible, innovative, and quick to change in order to capture any fleeting opportunities during a project.

Entrepreneurial

Micro-teams must be entrepreneurial in order to keep being successful. A single formula will not always be successful, and just as there is no one single way to complete a project, a leader must learn to keep being successful in difficult circumstances (Rigby, 2009).

Self-Organizing

A micro-team must be willing to change, resilient to external forces, and self-organizing. Teams must learn to act and react without direction (Duarte & Snyder, 2006; Lipnack & Stamps, 2000). Just as our immune systems respond to threats with various reactions, a self-organizing micro-team will take swift action based upon available resources and time.

VIRTUAL LEADERSHIP COMPLEXITY CHECKLIST

The virtual leadership checklist (Figure 19.3) offers a listing of details that virtual leaders should have in order to be successful. Leaders can use this checklist to review their personal skills, or it can be used as a guideline for certain important complexity-based leadership behaviors. The checklist can be given to individuals as a tool to have others rank their perspective regarding the leader.

An effective virtual leader who is applying complexity will answer all questions in the shaded area. Individuals who score outside the shaded areas should consider additional learning regarding complexity and leadership with regard to those selected areas. The figure will not assist the individual in the improvement of their skills (the self-assessment tool

Virtual Leadership Complexity Skills *Consider that an effective virtual leader with complexity skills will always answer in the shaded areas. Less effective leaders will be those that do not always answer in the shaded areas.*					
	Strongly Agree	Somewhat Agree	Neither agree nor disagree	Somewhat Disagree	Strongly Disagree
Trust - You trust your leader					
Trust - Your leader trusts you					
Trust - You trust the organization					
Integrity - Your leader displays integrity on a regular basis					
Integrity - Your leader is honest in all their actions					
Integrity - Your team is honest in all their actions					
Vision - Your leader has a vision of the future					
Vision - Your organization has a vision of the future					
Vision - Your community has a vision of the future that includes your organization					
Accountability - Your leader holds individuals accountable for their actions and inactions					
Accountability - The leader holds the team accountable for their actions or inactions					
Accountability - The leader holds external parties accountable for their actions or inactions with regards to the project					
Complexity - Your leader allows the team to be entrepreneurial					
Complexity - Your leader allows individuals to be entrepreneurial					
Complexity - Your leader allows the team to be self organizing with regards to process and policy					
Total					

Figure 19.3 Virtual Leadership Complexity Skills Checklist.

is for improvement) but it can offer some idea regarding an individual's current behavior and perspective on complexity.

If used as a tool the total section will assist in the scoring of the individual worksheets.

1. Total the number of answers in each column.
2. Multiply the total of the "Strongly Agree" responses by five.
3. Multiply the total of the "Somewhat Agree" responses by four.
4. Multiply the total of the "Neither Agree nor Disagree" responses by three.
5. Multiply the total of the "Somewhat Disagree" responses by two.
6. Multiply the total of the "Strongly Disagree" responses by one.

A score of 60 or more indicates that the individuals have achieved effective micro-teams or that the individual would be ready to serve on a micro-team. Scores of less than 60 should review the areas that were outside of the shaded areas in order to understand what room for improvement is available.

Virtual Leadership Complexity Self-Assessment

The virtual leadership self-assessment tool (Figure 19.4) offers individuals a method to see if they are currently expressing any virtual leadership skills that are applying complexity. It will rank the leader so that they understand how closely one is currently following the path of a complexity-based virtual leader. It will also help the project manager consider what areas are in need of further study or support while strengthening the current best practices. There is no right or wrong answer with the assessment, and it is only a tool to help improve the leader for the future.

Follow the directions on the self-assessment and complete the assessment in 30 to 45 minutes.

Once the self-assessment tool has been completed consider the following for Steps One to Four:

Every time the assessment did not prove or display the demonstration of the requisite virtual skill under consideration, rank that Step with a three.

Every time the assessment could prove or display that the requisite virtual skill under consideration, rank that Step with a four.

Name: Date:

Instructions	Expect to spend about 30-45 minutes with this self-assessment. You need to respond as accurately as possible and as fast as possible. Answer with the material that first comes to you rather then dwell too long on the question.

DETAILS:

STEP 1

Reflect upon your virtual leadership style and consider the following:

How do you instill trust?
How to your embody integrity?

STEP 2

Reflect upon your vision and consider the following:

How do you communicate accountability?
How do you display and communicate the vision to others?

STEP 3

Reflect upon your entrepreneurial spirit and consider the following:

How do you allow others to be entrepreneurial?
How do you communicate the entrepreneurial spirit to others?

STEP 4

Reflect if your team is self-organizing and consider the following:

What is keeping them from being more self-organizing?
What are you doing to help them become more self-organizing?

STEP 5

Reflect upon any of the elements of complexity that you are currently leveraging as a virtual leader and explain how you have communicated this learning to others.

Figure 19.4 Virtual Leadership Complexity Self-Assessment.

Every time the assessment could strongly or consistently display that the requisite virtual skill under consideration, rank that Step with a five.

For Step Five, score one point for every element of complexity that is either leveraged or communicated.

Total the score and then go to the scores interpretation section for more details.

Virtual Leadership Complexity Scores Interpretation

A score of 23 or more points means that the leader is already leveraging many elements of complexity in the virtual environment. The leader should review the self-assessment tool and consider any areas of opportunity for growth in the future. A leader who scores this high already has a strong understanding of complexity and is communicating this knowledge to others.

A score of 19 to 22 indicates that the leader has a good understanding of complexity in the virtual environment, with some areas for improvement. The leader should review the self-assessment tool and consider an area of opportunity and growth in the future. Any areas that were scored a three or less should be seriously reviewed in order to understand what is lacking. Consider reviewing the relevant sections of complexity in this book to see if new learning can be gleaned in order to improve in those areas.

A score of 18 or less indicates that the leader is not fully using complexity in virtual teams and there is considerable room for improvement in several areas. The leader should examine all of their answers and review the relevant areas in the text to determine what is currently lacking in their leadership style. A score of 18 or less does not mean that the individual is a poor leader; it only offers a relative scale of their application of complexity in virtual teams.

> Practical Tip: Virtual leadership and complexity should not be something that is forced quickly. Make sure that there are already elements of complexity as part of the organization before trying to increase complexity for greater benefit. Complexity is not a cure-all management panacea, but it is a new way to consider managing a project. It is designed to offer alternatives in different directions rather than solely relying upon linear system solutions.

COMPLEXITY TOOLS FOR BUILDING COMMUNITY

Building a Virtual Community with Complexity

Communities are all about communication and sharing information. A critical point is to communicate for people to buy into new ideas. In some ways this is like internal advertising, yet it is necessary for continued success (Kotter, 2002). Understanding how communities process data is critically important in levering this complexity. Figure 19.5 offers a sample virtual community that is organizationally connected but not colocated. Although the graphic explains the hierarchical connections between the different project team members, it does not explain how reciprocal the data exchange is. This graphic serves as a starting point for utilizing complexity.

Simply reviewing the formal relationships is not sufficient to understand how communities communicate. It is necessary to review and map this communication at the individual level. This will allow a complete

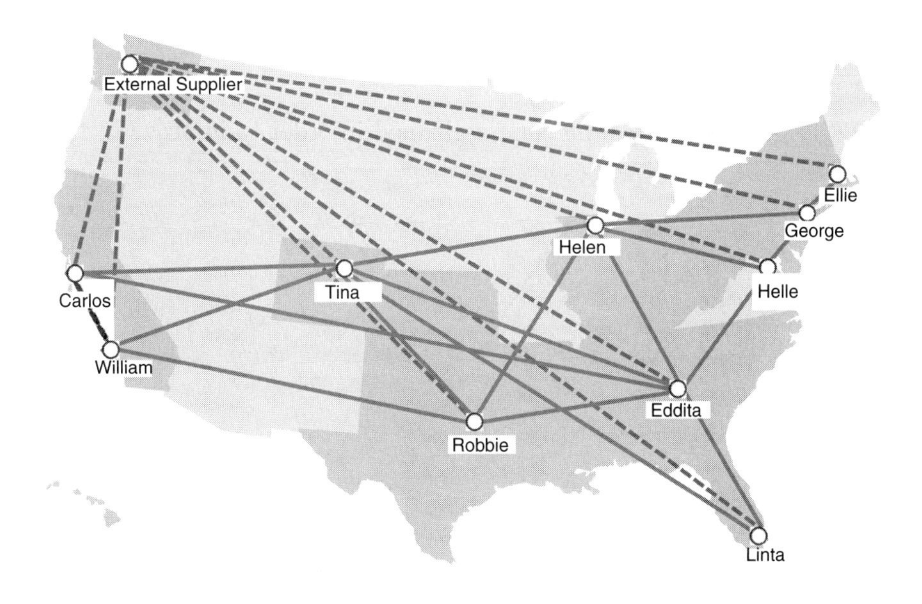

Figure 19.5 Sample Virtual Community.

understanding of how and through what method every individual communicates with one another.

Community Communication Mapping for Understanding Complexity

The community communication mapping offers a direct method to understanding how complex communication travels through an organization. The basic tool (Figure 19.6) shows how to build the community.

One should map everyone in the project team in order to see how they interact with one another and how with the community. Once every team member is mapped (including yourself), combine the maps to see if there is any additional touchpoints. Finally, compare this mapping to an

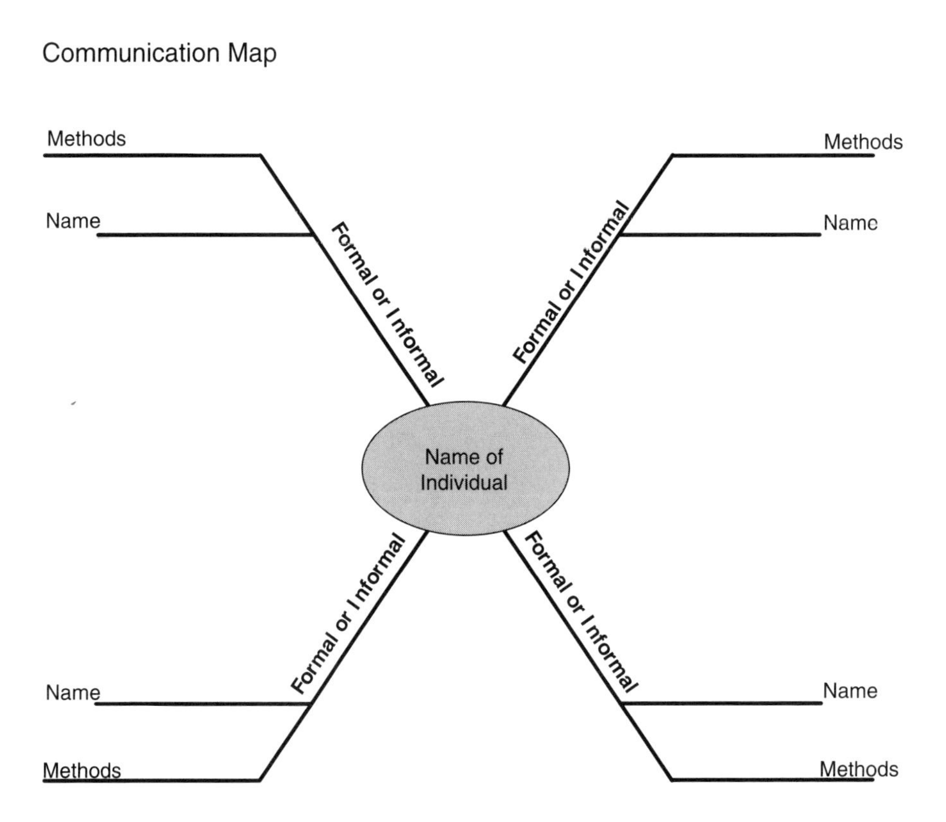

Figure 19.6 Individual Communication Map.

organizational chart to see how the formal and informal communication works together. Were any unrealized connections discovered from this process? Give those touchpoints greater consideration and reflect upon what other new information has been learned from this process.

When this is all complete, one should then be able to understand how the project team interacts with the community and how information is distributed through the organization. Once this communication network is understood then one is able to leverage information in different media in order to spread news throughout the community.

Complexity Checklist for Building Community

The self-assessment tool offers a method to see if the individual is currently expressing any of the complexity skills with regard to the community (Figure 19.7). It will rank the individual in a way to understand how closely one is leveraging skills associated with complexity and community. It will also help the individual consider what areas are in need of further study or support while strengthening the current best practices. There is no right or wrong answer with the assessment, and it is only a tool to help improve the individual for the future.

Follow the directions on the self-assessment and complete the assessment in 20 to 30 minutes.

Once the self-assessment tool has been completed, consider the following for Steps One to Four:

Every time the assessment did not prove or display the demonstration of the requisite virtual skill under consideration, rank that step with a three.

Every time the assessment could prove or display that the requisite virtual skill under consideration, rank that step with a four.

Every time the assessment could strongly or consistently display that the requisite virtual skill under consideration, rank that step with a five.

For Step Five, score one point for every element of complexity that is either leveraged or communicated.

Total the score and then go to the scores interpretation section for more details.

Name: Date:

Instructions	Expect to spend about 20-30 minutes with this self-assessment. You need to respond as accurately as possible and as fast as possible. Answer with the material that first comes to you rather then dwell too long on the question.

DETAILS:

STEP 1

Reflect upon complexity and community and consider the following:

Do you regularly communicate with your team?

STEP 2

Reflect upon complexity and community and consider the following:

How efficiently do you communicate?

STEP 3

Reflect upon complexity and community and consider the following:

Describe how effective you are with multiple means of communication?

STEP 4

Reflect if your team is connected to the community and consider the following:

What keeps them connected to the community?
What is being done to help them be more connected to the community?

STEP 5

Reflect upon any other elements of community, complexity and communication that is currently being leveraging and explain how this is being shared with others.

Figure 19.7 Community Complexity Self-Assessments.

Community Complexity Scores Interpretation

A score of 23 or more points means that the individual is already leveraging many elements of complexity and communities. The individual should review the self-assessment tool and consider any areas of opportunity for growth in the future. An individual who scores this high already has a strong understanding of complexity and is communicating this knowledge to others.

A score of 19 to 22 indicates that the leader has a good understanding of complexity and communities and has some areas for improvement. The individual should review the self-assessment tool and consider an area of opportunity and growth in the future. Any areas that were scored a three or less should be seriously reviewed in order to understand what is lacking. Consider reviewing the relevant sections of community and complexity in this book to see if new learning can be garnered in order to improve in those areas.

A score of 18 or less indicates that the leader is not fully using complexity and communities and there is considerable room for improvement in several areas. Individuals should examine all their answers and review the relevant areas in the text to determine what is currently lacking in their knowledge of complexity and community. A score of 18 or less does not mean that someone is ineffective in their community; it only offers a relative scale of the individual's application of complexity in communities.

COMPLEXITY TOOLS FOR BUILDING TEAMS

Building Team Diagram

Teams are an important part of complexity and understanding how they interact is important for the application of complexity (Figure 19.8). The figure offers a clear view of what elements are parts of a complexity-based team as well as explaining the important interactions of these important factors. This graphic will serve to offer a perspective on teams and complexity in order to assist with how this fits inside a project.

Four important elements of teams are trust, vision, accountability, and motivation. All of these elements are required in order to build successful teams. Project managers must do their best to cultivate the four factors in order to create teams that will not only succeed but thrive in difficult projects.

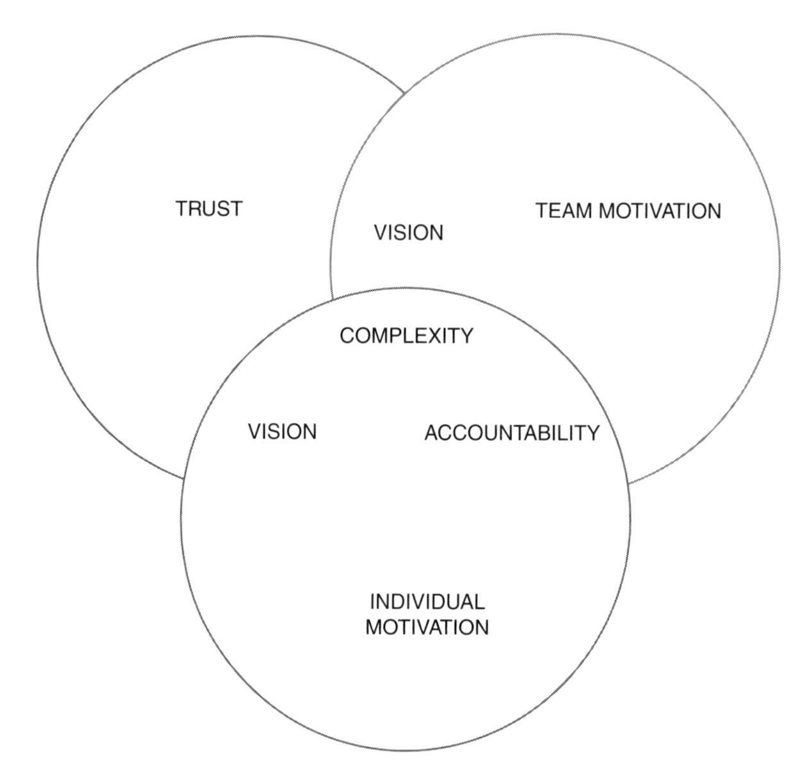

Figure 19.8 Teams in Complexity.

Complexity Checklist for Building Teams

The complexity checklist offers a listing of elements necessary for a virtual team that leverages complexity (Figure 19.9).

An effective team that applies complexity will answer all of the checklist questions in the shaded area. Individuals that score outside of the shaded areas should consider additional learning regarding teams and complexity with regards to those selected areas. The checklist will not assist the individual in the improvement of their team skills (the self-assessment tool is for improvement) but it can offer some idea regarding an individual's current behavior and perspective on team complexity.

If used as a tool the total section will assist in the scoring of the individual worksheets.

1. Total the number of answers in each column.
2. Multiply the total of the "Strongly Agree" responses by five.
3. Multiply the total of the "Somewhat Agree" responses by four.

Consider that an effective team leader or member with complexity skills will always answer in the shaded areas. Less effective team leaders or members will be those that do not always answer in the shaded areas.					
	Strongly Agree	**Somewhat Agree**	**Neither agree nor disagree**	**Somewhat Disagree**	**Strongly Disagree**
Trust - Do you trust your leader					
Trust - Your leader trusts you					
Trust - You trust the organization					
Vision - You have a vision of the future					
Vision - Your organization has a vision of the future					
Vision - Your community has a vision of the future that includes your organization					
Accountability - You hold yourself accountable for your actions and inactions					
Accountability - You hold other team members accountable for their actions or inactions					
Accountability - You hold external parties accountable for their actions or inactions with regards to the project					
Motivation - You actively motivate yourself to achieve more					
Motivation - You actively motivate the team to achieve more					
Motivation - You actively motivate external individuals to achieve more for the project					
Complexity - You are entrepreneurial					
Complexity - Your teams is entrepreneurial					
Complexity - Your external stakeholders are entrepreneurial					
Total					

Figure 19.9 Team Complexity Skills.

4. Multiply the total of the “Neither Agree nor Disagree” responses by three.
5. Multiply the total of the “Somewhat Disagree” responses by two.
6. Multiply the total of the “Strongly Disagree” responses by one.

A score of 60 or more indicates that the individuals are utilizing complexity for effective teams or that the individual would be ready to be on a complexity-based team. Scores of less than 60 should review the areas that were outside the shaded areas in order to understand what room for improvement is available.

Complexity Self-Assessment for Building Teams

The self-assessment tool offers a method for individuals to determine if they are ready to build a team that can support complexity (Figure 19.10). The self-assessment tool will rank the individual in order to understand how effective the person would be with a complexity-based team. It will also help the project manager consider what areas are in need of further study or support while strengthening the current best practices. There is no right or wrong answer with the assessment, and it is only a tool to help improve the individual for complexity-based teams in the future.

Follow the directions on the self-assessment and complete the assessment in 30–45 minutes.

Once the self-assessment tool has been completed, consider the following for Steps One to Four:

Every time the assessment did not prove or display the demonstration of the requisite team skill under consideration, rank that step with a three.

Every time the assessment could prove or display that the requisite team skill under consideration, rank that step with a four.

Every time the assessment could strongly or consistently display that the requisite team skill under consideration, rank that step with a five.

For Step Five, score one point for every element of complexity that is either leveraged or communicated.

Total the score and then go to the scores interpretation section for more details.

Name: Date:

Instructions	Expect to spend about 30-45 minutes with this self-assessment. You need to respond as accurately as possible and as fast as possible. Answer with the material that first comes to you, rather then dwell too long on the question.

DETAILS:

STEP 1

Reflect upon your team experience and consider the following:

How do you instill trust?
How to your embody integrity?

STEP 2

Reflect upon your vision and consider the following:

How do you communicate accountability?
How do you display and communicate the vision to others?

STEP 3

Reflect upon your ability to hold yourself and others accountable and consider the following:

How do you hold yourself accountable in difficult situations?
Do you hold others accountable?

STEP 4

Reflect if your team is self-motivating and consider the following:

Are you self-motivating?
Are team members self-motivating?

STEP 5

Reflect upon any of the elements of team complexity that you are currently leveraging and explain how you have communicated this information to others.

Figure 19.10 Complexity Self-Assessment for Teams.

Complexity-Based Teams Scores Interpretation
A score of 23 or more points means that the individual is already leveraging many elements of teams and complexity. The individual should review the self-assessment tool and consider any areas of opportunity for growth in the future. An individual who scores this high already has a strong understanding of complexity-based teams and is communicating this knowledge to others.

A score of 19 to 22 indicates that the individual has a good understanding of complexity based teams but has some areas that need improvement. The individual should review the self-assessment tool and consider any area of opportunity and growth in the future. Any areas that were scored a three or less should be seriously reviewed in order to understand what is lacking. Consider reviewing the relevant sections of complexity in this book to see if new learning can be gleaned in order to improve in those areas.

A score of 18 or less indicates that the individual is not fully using complexity in teams and there is considerable room for improvement in several areas. The individual should examine all of answers and review the relevant areas in the text to determine what is currently lacking in understanding and style of micro-team building. A score of 18 or less does not mean that the individual is poor at working in teams; it only offers a relative scale of the person's application of complexity in teams.

COMPLEXITY TOOLS FOR BUILDING MICRO-TEAMS (TRIBES)

Building Micro-Team Diagram

Figure 19.11 offers a view to how a micro-team fits within the realm of complexity and project management. The figure offers a clear view of what elements are part of a complexity-based micro-teams and what elements are better served by the *PMBOK*® (PMI, 2008). This view illustrates what should be considered as part of complexity and what should be considered as more of traditional project management. It is imperative that one understands the difference to avoid applying the wrong solution to a given situation. This does not mean that complexity-based solutions are not effective in all cases; it just offers a perspective to avoid using complexity for all possible solutions when other tools are available.

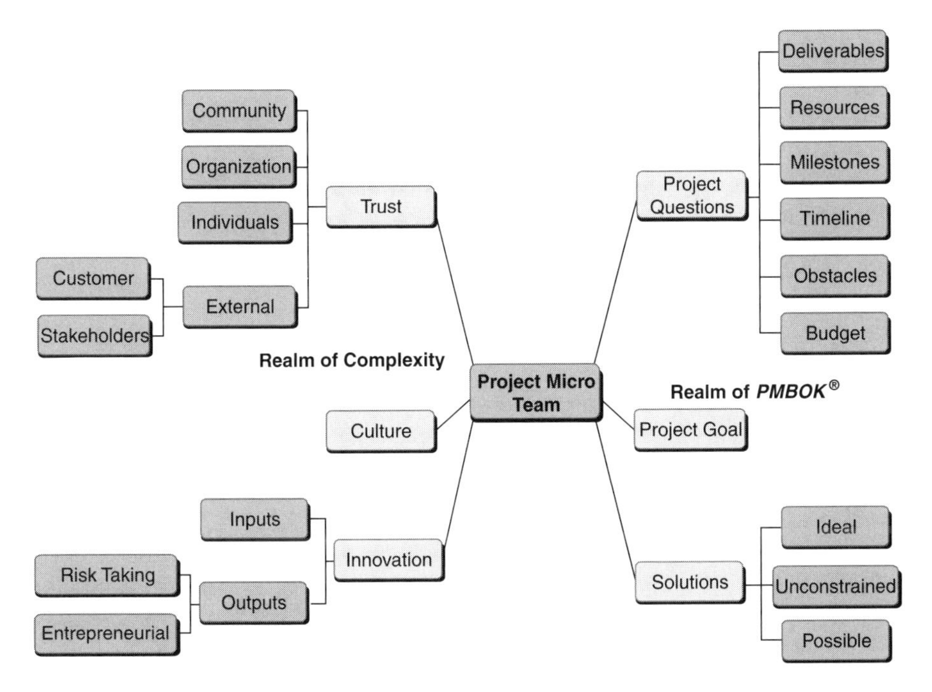

Figure 19.11 Complexity of micro-teams.

There are three important elements to all micro-teams: trust, culture, and innovation. All of these elements are required in order to build successful micro-teams. Project leaders must do their best to cultivate all three of these elements in order to create teams that will not only succeed but thrive in difficult projects.

Trust

Trust in a micro-team is more critical as there is no room for having others pick up the slack (Hamel, 2009). Everyone on the micro-team must trust everyone to the extreme (Godin, 2008). The reciprocal relationship between the leader and the follower must be even tighter than in a larger group as the perception will be that the leader should have more time for personalize attention. This is not always the case but this is the subliminal perspective. There must also be considerable mutual respect and understanding.

Culture

The culture of a micro-team is very important because of the close-knit relationship of all the team members (Godin, 2008). A micro-team is also very susceptible to the external cultural forces because of the small size of the group. If one person begins to display new cultural behaviors, then it is very likely that others will follow suit because of the extreme influence that team members have upon one another. Undue peer pressure can influence the micro-team to develop new culture elements, so one must remain careful about what cultural elements are adopted by the group.

Innovation

An advantage of the micro-team is that it can be very nimble (Godin, 2008), can change rapidly, and can make adjustments quickly. A micro-team should be given the latitude to be highly innovative. A small group that is allowed to make rapid changes can capture short-term opportunities in a project or in the marketplace. It can also respond quickly when a new innovation fails to work or when another innovation is no longer as effective. Complexity can help drive this innovation; more interest is generated toward innovation, the more innovative groups can become.

Complexity Checklist for Building Micro-Teams

The complexity checklist offers a listing of elements that are necessary for a virtual micro-team that leverages complexity (Figure 19.12).

An effective micro-team that is applying complexity will answer all questions in the shaded area. Individuals who score outside of the shaded areas should consider additional learning regarding micro-teams related to the selected areas. The checklist will not assist the individual in improving micro-team skills (the self-assessment tool is for improvement), but it can offer some idea regarding an individual's current behavior and perspective on micro-team complexity.

If used as a tool, the total section will assist in the scoring of the individual worksheets.

7. Total the number of answers in each column.
8. Multiply the total of the "Strongly Agree" responses by five.
9. Multiply the total of the "Somewhat Agree" responses by four.
10. Multiply the total of the "Neither Agree nor Disagree" responses by three.

Consider that an effective micro-team with complexity skills will always answer in the shaded areas. Less effective micro-teams will be those that do not always answer in the shaded areas.

	Strongly Agree	Somewhat Agree	Neither agree nor disagree	Somewhat Disagree	Strongly Disagree
Trust - You trust your leader					
Trust - Your leader trusts you					
Trust - You trust the organization					
Culture - Your leader displays the organizational values on a regular basis					
Culture - Your team displays the organizational values on a regular basis					
Culture - Your community recognizes your organizational values					
Innovation - You are allowed to be innovative					
Innovation - Your leader is innovative					
Innovation - Your community recognizes that your team is innovative					
Complexity - Your leader allows the team to be entrepreneurial					
Complexity - Your leader allows individuals to be entrepreneurial					
Complexity - Your leader allows the team to be self organizing with regards to process and policy					
Total					

Figure 19.12 Building Micro-Teams with Complexity Skills.

11. Multiply the total of the "Somewhat Disagree" responses by two.
12. Multiply the total of the "Strongly Disagree" responses by one.

A score of 48 or more indicates that the individuals have achieved effective micro-teams or that the individual would be ready to serve on a micro-team. Scores of less than 48 should review the areas that were outside of the shaded areas in order to understand what room for improvement is available.

Complexity Self-Assessment for Building Micro-Teams

The complexity self-assessment tool for the creation of micro-teams offers an individual a method to reflect upon how the individual is currently expressing any micro-team skills that are applying complexity. It will rank the individual so they understand how effective the person would be with micro-teams. It will also help the project manager consider what areas are in need of further study or support while strengthening the current best practices. There is no right or wrong answer with the assessment; it is only a tool to help improve the leader for the future.

The self-assessment tool offers individuals a method to determine if they are ready to build a virtual micro-team that can support complexity (Figure 19.13). Follow the directions on the self-assessment and complete the assessment in 30–45 minutes.

Once the self-assessment tool has been completed, consider the following for Steps One to Four:

Every time the assessment did not prove or display the demonstration of the requisite virtual skill under consideration, rank that step with a three.

Every time the assessment could prove or display that the requisite virtual skill under consideration, rank that step with a four.

Every time the assessment could strongly or consistently display that the requisite virtual skill under consideration, rank that step with a five.

For Step Five, score one point for every element of complexity that is either leveraged or communicated.

Total the score and then go to the scores interpretation section for more details.

Complexity-Based Micro Teams Scores Interpretation

A score of 23 or more points means that the individual is already leveraging many elements of micro-teams and complexity. The individual should review the self-assessment tool and consider any areas of opportunity for growth in the future. An individual who scores this high already has a strong understanding of complexity-based micro-teams and is communicating this knowledge to others.

A score of 19 to 22 indicates that the leader has a good understanding of complexity-based micro-teams but some areas need improvement. The individual should review the self-assessment tool and consider an area of opportunity and growth in the future. Any areas that were scored a

Name: Date:

Instructions	Expect to spend about 30-45 minutes with this self-assessment. You need to respond as accurately as possible and as fast as possible. Answer with the material that first comes to you, rather then dwell too long on the question.

DETAILS:

STEP 1

Reflect upon your current team (or micro-team) and consider the following:

How do you instill trust?
How to your embody integrity?

STEP 2

Reflect upon your current organizational culture and consider the following:

How do you communicate the culture of the organization?
How do you display and communicate the culture to others?

STEP 3

Reflect upon your innovative ability and consider the following:

How do you allow others to be innovative?
How do you infect others with an innovative spirit?

STEP 4

Reflect if your team is self-organizing and consider the following:

What is keeping them from being more self-organizing?
What are you doing to help them become more self-organizing?

STEP 5

Reflect upon any of the elements of complexity that you are currently leveraging in your current team and explain how you have communicated this information to others.

Figure 19.13 Complexity Self-Assessment for Building Micro-Teams.

three or less should be seriously reviewed in order to understand what is lacking. Consider reviewing the relevant sections of complexity in this book to see if new learning can be gleaned in order to improve in those areas.

A score of 18 or less indicates that the individual is not fully using complexity in micro-teams and there is considerable room for improvement in several areas. The individual should examine all their answers and review the relevant areas in the text to determine what is currently lacking in the understanding and style of micro-team building. A score of 18 or less does not mean that the individual is poor at working in teams; it only offers a relative scale of the application of complexity in micro-teams.

COMPLEXITY TOOLS FOR DEALING WITH CHANGE

Tools for Dealing with Change while Moving toward Complexity

Change is a constant in life. There may have been a time long ago where things remained relatively static and people were rewarded for resisting change. But the world has moved past that time and everyone must learn to deal with the rapidly changing world. The change checklist (Figure 19.14) should be a reminder that everyone must deal with change.

Complexity embraces change as complexity theory recognizes that with all change comes some chaos and some order (Pievani & Varchetta, 2005). Both are required to live in the future. Consider using copies of the change checklist on a daily basis for thirty days. Use this simple checklist to track your daily tasks. Keep copies of all the completed checklists and then review the accomplishments after thirty days.

Review what change has occurred from day one to day thirty. You may be surprised at how much change happens when you put it at the top of the agenda every day. Count the number of days that you were able to check off the change box at the top of your to-do list after thirty days. Reflect upon your findings and consider how much complexity is entering your life.

Leveraging the Complexity of Language for Dealing with Change

One of the most robust elements of complexity is language. Human languages are probably some of the most complex systems on the planet and

Things To Do ☑

☐ **CHANGE** ____________________

☐ ____________________

☐ ____________________

☐ ____________________

☐ ____________________

☐ ____________________

☐ ____________________

☐ ____________________

☐ ____________________

☐ ____________________

☐ ____________________

☐ ____________________

☐ ____________________

☐ ____________________

☐ ____________________

☐ ____________________

☐ ____________________

☐ ____________________

☐ ____________________

Figure 19.14 Change Checklist.

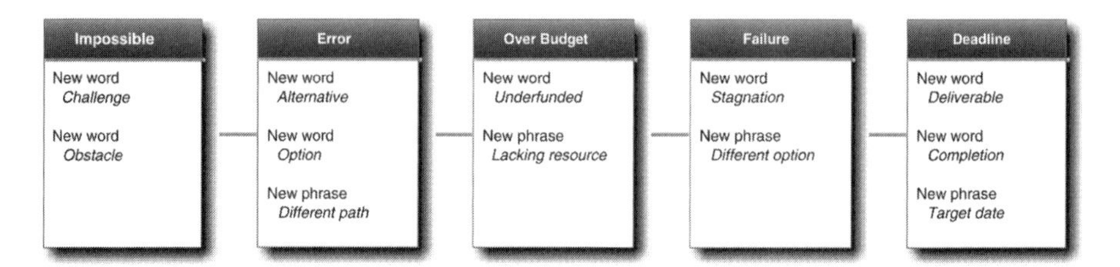

Figure 19.15 New Language of Change.

there are many different languages in use today. Since change is disconcerting to everyone, it is often a good idea to try to make people more comfortable with change on a daily basis. One complexity-based solution is to leverage language in pursuit of change. Not only is this method effective, but it is free.

Consider Figure 19.15, which shows how there are many related and similar words available for use, instead of using certain words that carry more negative connotations. This listing is just an example because there are hundreds of other words that could be replaced in order to assist with change.

Create a personal list of words related to change and then get into the habit of using them. You can even try to correct others who use the old word in order to better spread the use of the new words. This method can truly change the organization as over time more and more people will use the new words and change will become less negative.

This may seem simple or perhaps even a little odd, but the method is effective. Words carry multiple meanings and people will be able to accept change better if there is not a sense of pending doom associated with the change. Keep in mind that most people dread more in advance of change than from the actual change, so using words in advance can help ease the transition. Watch the change happen over time—it is surprising.

Complexity Self-Assessment for Dealing with Change

The complexity self-assessment tool for the acceptance of change offers individuals a method to reflect upon how they are currently coping with change. The self-assessment tool will rank someone in order to offer some perspective on how well the individual deals with change. It will also help the individual consider what areas are in need of further study or support while strengthening the current best practices. There is no right or wrong answer with the assessment; it is only a tool to help prepare the individual for change in the future.

The self-assessment tool offers a method for individuals to determine if they are ready to accept change and complexity (Figure 19.16). Follow the directions on the self-assessment and complete the assessment in 30–45 minutes.

Once the self-assessment tool has been completed, consider the following for Steps One to Four:

Every time the assessment did not prove or display the demonstration of the requisite skill under consideration, rank that step with a three.

Every time the assessment could prove or display the requisite skill under consideration, rank that step with a four.

Every time the assessment could strongly or consistently display the requisite skill under consideration, rank that step with a five.

For Step Five, score one point for every element of complexity that is either leveraged or communicated.

Total the score and then go to the scores interpretation section for more details.

Complexity-Based Micro Teams Scores Interpretation

A score of 23 or more points means that the individual is already leveraging many elements of change and complexity. The individual should review the self-assessment tool and consider any areas of opportunity for growth in the future. An individual who scores this high already has a strong understanding of complexity-based change and is communicating this knowledge to others.

A score of 19 to 22 indicates that the individual has a good understanding of complexity-based communication and change but has some areas of improvement. The individual should review the self-assessment tool and consider any areas of opportunity available to them in the future. Any areas that were scored a three or less should be seriously reviewed in order to understand what skill the individual is lacking. Consider reviewing the relevant sections of complexity in this book to see if new learning can be gathered in order to improve in those areas.

A score of 18 or less indicates that the individual is not fully using complexity and change and there is considerable room for improvement in several areas. These individuals should examine all of their answers and review the relevant areas in this book to determine what is currently lacking in their understanding and style of coping with change and

Name: Date:

Instructions	Expect to spend about 30-45 minutes with this self-assessment. You need to respond as accurately as possible and as fast as possible. Answer with the material that first comes to you, rather then dwell too long on the question.

DETAILS:

STEP 1

Reflect upon your acceptance of change and consider the following:

How do you cope with change?
How do you show others how to cope with change?

STEP 2

Reflect upon your understanding of change and consider the following:

How do you communicate change to others?
How do you communicate change acceptance to others?

STEP 3

Reflect upon how you resist change and consider the following:

How do you keep from resisting change?
How do you keep others from resisting change?

STEP 4

Reflect upon when you accept change and consider the following:

What helps you accept change?
How do you help others to accept change?

STEP 5

Reflect upon any of the elements of change and complexity that you are currently leveraging and explain how you have communicated this experience to others.

Figure 19.16 Change and Complexity Self-Assessments.

complexity. A score of 18 or less does not mean that an individual is poor at handling change; it only offers a relative scale of their application of complexity and organizational change.

CHAPTER SUMMARY

Complexity virtual projects are some of the hardest projects to manage. The more that can be done to help keep the project focused, the more successful the project will be. Leadership, communities, teams, tribes, and change are all aspects of a virtual project. The more effective an individual can be in these circumstances, the more successful they will be.

Consider using at least one of these tools in your next project and recommending at least one of these tools to your project manager to improve the project. It may not help the project this time, but it will help you become familiar with them so that they will be more effective in the next project. Tools are best when they are used consistently and regularly, so make the best of these and continue to learn from what happens when they are used.

REFERENCES

Bass, B. (1990). Bass & Stogdill's Handbook of leadership: Theory, research, & managerial applications. (3rd ed.). New York: The Free Press.

Bennis, W. (1994). On becoming a leader. New York: Addison-Wesley.

Duarte, D., & Snyder, N. (2006). *Mastering virtual teams: Strategies, tools, and techniques that succeed* (3rd ed.). San Francisco: Jossey-Bass.

Godin, S. (2008). Tribes. NY, NY: Penguin Group.

Giblin, E. J., & Amusco, L. E. (1996). Putting meaning into corporate values. Business Forum, *22*(1), 14–18.

Hamel, G. (2007). The Future of Management. Boston, MA: Harvard Business School Press.

Lipnack, J., & Stamps, J. (2000). Virtual Teams: People working across boundaries with technology (2nd ed.). New York: John Wiley & Sons.

Jaafari, A. (2003). Project management in the age of complexity and change. *Project Management Journal*, *34*(4), 47–57.

Kotter, J. P. (2002). The heart of change: Real-life stories of how people change their organizations. Boston: Harvard Business School Press.

Project Management Institute (Ed.). (2008). *A Guide to the Project Management Body of Knowledge—Fourth Edition*. Newtown Square, PA: PMI.

Pievani, T. & Varchetta, G. (2005). The strategies of uniqueness: Complexity, evolution, and creativity in the new management theories . . . or, in other words, what is the connection between an immune system network and a corporation. World Futures, 61. Milan, Italy: Routledge Taylor Francis Group.

Rigby, D. (2009). Winning in turbulence. Boston, MA: Harvard Business Press.

Tichy, N. (2004). The cycle of leadership: How great leaders teach their companies to win. NY, NY: HarperBusiness.

Chapter 20

Virtual Projects and Complexity Theory

Figure 20.1 Consider a virtual project as the river in the picture and complexity theory as the bridge that spans this divide. Virtual projects are challenges that need to be crossed and complexity is a method to cross those challenges.

COMPLEXITY THEORY AS APPLIED TO SUCCESSFUL VIRTUAL LEADERS

One of the essential applications of complexity theory is the movement away from the underlying leadership maxim that there is just one best way to lead people. In the past, there was an attempt to find the one right way to manage people. This assumption was rooted in management theory for a very long time and whenever a theory came forward that did not cover all the possibilities, it would be rejected and then the next management theory would come forward. Over time more and more management theories came forward, and often they would reflect society, technology, and business. As leadership evolved, more ideas were presented and rejected (Bass, 1990).

Just as society has changed, leadership has changed in a manner to reflect these sociological changes. Leadership theory became a jumbled mass of completing and conflicting ideas, all with research to support their findings, but none that was the one true path. As society tried to digest this information, many of the applicable, successful, and sometimes best-marketed leadership paradigms began to shape the view about leadership.

Some research has tried to categorize leadership into different schools of thought, such as *old school* and *new school*. This kind of division creates a feeling that there was a split in thinking and is based upon one simple rule of leadership theory of the past and present. There has been a rift in leadership thinking, which now holds that there may be multiple best solutions to leadership, but even that answer is somewhat dissatisfying. The reason is that if there are multiple options, there will always be a debate on which one is best.

Virtual leaders have already seen that transformational leadership is one that is more successful than other forms of leadership. Complexity theory moves from the thinking that the leader is the font of all knowledge and there is one right way of leadership, to the thinking that leaders must convert followers into leaders. Other leadership theories dwell on the fact that the style, structure, or format of the leader's style is what is important and that is what drives leadership. Just as complexity theory moves away from trying to explain all complex systems by examining a few variables, transformational leadership moves away from the leader-centric concept of leadership. It is true that the leader is important, but not as the only active variable in the system. Transformational leadership sees the leader as a dynamic partner in the dance of virtual organizational leadership.

Transformational leaders transform followers into leaders. In a 1984 study, Warren Bennis (b. 1925) was able to identify 90 transformational leaders. He found evidence to support the following elements of leadership: competence to manage attention and meaning, the communication of a possible vision, and the empowerment of those working toward a collective goal (Bennis, 1995). Transformational leadership is a set of skills and behaviors capable of being learned and managed. It is a leadership process that is systematic, consisting of purposeful and organized search for changes, systematic analysis, and the capacity to move resources from areas of lesser to greater productivity.

> Practical Tip: Reflect upon your leadership style and consider if you operate from a theory of contingency leadership. If you do, consider what you can do to become more of a transformational leader. If you already are a transformational leader, consider what elements of contingency leadership might be helpful to your leadership style.

A transformational leader is one who is not only versed in the scope and deliverables of a project and the goals and objectives of the project but must also understand the goals and objectives of the project team, the project stakeholders, and the organizations involved the project. To this end, the transformational leader can understand some of the normally hidden dynamics of organizational decision making. A transformational leader has a base of this information but understands the deliverables and goals of the project in a manner that can be communicated to others. Although many of these elements that have been attributed to the transformational leader are addressed in the *PMBOK*®, the process is seen as one of inputs and outputs with a communication plan and reporting structure. This is essentially a static plan where the goal of the project manager is to herd the individuals back to the chosen path if people stray away from the plan.

A transformational leader can change direction and create a new plan in order to achieve the same goal. Leaders should not need to make an entirely new project plan every time there is an environmental change, but transformational leaders need to be willing to reinvent themselves, their projects, and their project objectives in order to meet the needs of the project and the project team.

APPLICATION OF COMPLEXITY THEORY TO SUCCESSFUL VIRTUAL TEAMS

There is a fine line between chaos and keeping order on a virtual project. A project manager must decide if he or she is to become an incessant micromanager or be able to create micro-teams to deal with small crises when they arise. The VPM has to ensure that there is just enough order to ensure forward momentum but not so much resistance that will stifle self-organization and creativity. To this end, culture becomes the roadmap for small team success. Small teams need general guidelines in order to remain focused. It is clear that the project manager needs to create the vision, establish an environment of trust, provide the tools through which the teams can communicate and collaborate freely, and most importantly make sure that the project manager can step back and allow the project team to function independently with minimal oversight (Lucey, 2008). The project manager must step away from the day-to-day hand holding and become the stitching that holds together the fabric of the project. Furthermore, leaders must lead from the front and be visible to make a positive impact upon their teams (Marcinko & Weisman, 1997). This visibility is even more important in the beginning and if the project runs into difficult circumstances.

Projects in the business world have become more complex and will rely upon fewer and fewer individuals to produce greater and greater results. On the technical/mathematical side of academics, with the advent of computers there arose complexity/chaos theory. In short, the complexity theory simply stated is there are systems too complex to define but appear to have patterns with some meaning. As stated before, the project manager must keep in mind certain previously discussed tangential theories that form the basis of complexity theory, such as the butterfly effect and six degrees of separation (Cooke-Davies, Cicmil, Crawford, & Richardson, 2007; Singh & Singh, 2002).

A successful strategy is for the virtual project manager to calculate the formal and informal lines of communication (LOC). Understanding how information flows within an organization, even on a simple virtual project, makes for an important exercise for the project manager. Keep in mind that there will always be formal, hierarchical lines of communication and informal, non-hierarchical lines of communication. It is best to map out these lines in order to better understand how information is disseminated within an organization. Furthermore, it helps a virtual project manager understand how to better communicate with their organization.

Figure 20.2 offers a simple mapping of communication within a project. As we can see, the virtual project manager communicates formally to all the members of the team, but it is quickly apparent that information sharing is occurring informally within the team. The diagram quickly shows that James is the one degree of separation communicator to the rest of the team. There is a path from Joe to Julie; that information must always pass to James and he must then pass on that message. Keep in

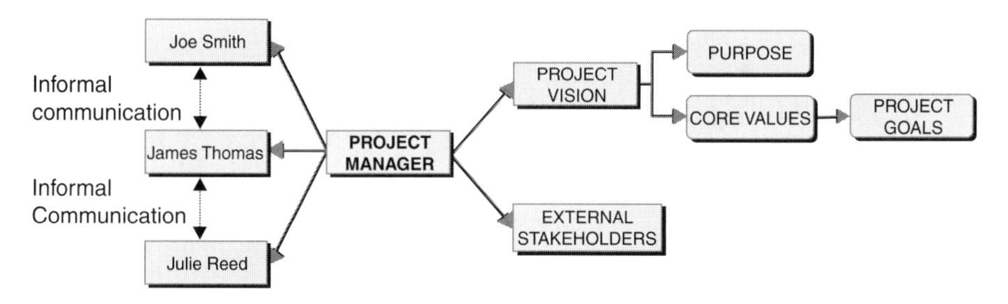

Figure 20.2 Organizational Communication Mapping.

mind that the more distance the message must travel, the fainter the signal and the more prone the message is to being misrepresented. Think about playing the game of telephone: a message is whispered to one child who then passes that message along the line. The message the last child receives is nothing like the message that was first whispered.

Another application to a virtual project would be to think about complexity as a way to break down all the parts of a virtual project and view the project from the lens of a master clockmaker. Keep in mind that each element interacts in a way to achieve a larger goal, but the examination of each element might not be a miniature version of the bigger picture. For example, before the establishment of the United States, England controlled its colonies through a series of autonomous governors appointed by the crown. These governors had great latitude in their implementation of their rule, and as long as the crown continued to receive what it believed to be its fair share of taxes, the governors were left to handle the day-to-day business. Using this metaphor, one can reflect upon how a project manager who oversees a large multinational project might be successful.

Trust and accountability are central to allowing monitored complexity to go forward (Jarvenpaa, Shaw, & Staples, 2004). There is some structure that is seen in complexity. The VPM needs to have the standard processes and procedures as defined by the *PMBOK® Guide* (PMI, 2008) or through a common methodology like PRINCE 2 or some other defining requirement. However, the processes and procedures should allow for self-organization, sharing of power, creating trust, flexibility to adapt as necessary, and most importantly a goal.

When there is a disruption to a team on the project, the project manager may or may not be aware of it. The project manager cannot automatically insert himself or herself to resolve it unless asked. Let the team resolve it. The team will be more adept at resolving conflict if allowed to resolve it on its own. In fact, depending on when it happens, it may be resolved before you even find out that there is a problem. If the project manager is always there to correct the problem, the team will learn to wait rather than to take corrective action. Probably the hardest part of embracing complexity theory is letting go. Too often the project manager clings to what they know and what they know is intervention rather than allowing for a good system to generate good results.

A futurist once described that the manufacturing plant of the future would consist of the factory, one man, and one dog. The man was there to feed the dog and the dog was there to keep the man from tinkering with the machine. It would appear that complexity theory is more about the project manager letting go and letting a perfect system develop rather than continually intervening in the belief that their sole role is to make sure that things are accomplished.

CREATING CONTINGENCIES THAT APPLY COMPLEXITY THEORY

Complexity theory is apt for a virtual project. A well-run virtual project verges on controlled chaos since there are many lines of communication and lack of visual cues. A project manager who has never met the team leads or the majority of the project team members many times leads the virtual project. This project manager is then supposed to bring this project in on time and within budget and within scope/quality. The lines of communication (n(2n/2+n-1)) can become daunting especially since there is the added complexity of no body language, which may account for 70 percent of the message (Roebuck, 2001). Understanding that a virtual project may result in more chaos, it is wise for a project manager to have certain contingencies in order to maintain the balance of control. A virtual project manager should have contingencies for communication, for the culture, and for the project.

A communication contingency would be how the virtual project manager will handle any disruptions in the flow or dissemination of communication. The virtual project manager should consider either creating redundant systems or having additional methods of communication. This should not be confused with having a technology contingency in case the phones or e-mails fail; it concerns how the team communicates in the event of a disruption in the organization, the team, or the culture. The project manager needs to understand how the organization and the team communicate in order to make sure that information continues to flow out to everyone involved. This may mean that the project manager makes personal contact with everyone involved very briefly in order to bring them up to speed. It may also setting up town hall meetings to allow everyone to attend and gain insight to the project progress. Regardless of the manner of contingency communication, the project manager needs to figure out a backup method and utilize it in case there is a problem with the typical communication.

This communication contingency also becomes important if the project has problems. The project manager should immediately implement the communication contingency plan in order to ensure that the message is getting out about the project. Making sure that all stakeholders are aware of a project meeting challenges is critically important.

A cultural contingency is important to the virtual project manager so that they understand how to move the culture away from potentially negative issues. For example, how will the project manager help a team move past a particularly problematic individual? A project may have a problematic sponsor who is highly negative. This kind of negativity can bring the team into a negative mode that is difficult to climb out of. The project manager should understand the culture well enough in order to move individuals past these kinds of problems.

This may mean that the project manager has a number of positive celebrations in order to move everyone out of a negative view of the project. It may mean breaking up the team responsibilities in order to give people a new perspective on matters. Often when team members get a better view of other people's positions and challenges they become more helpful and positive moving forward. A cultural contingency needs to shift the entire community to a more positive vantage point. The virtual project manager should consider keeping this positive message in the forefront of people's minds in order to get past this hurdle.

Project contingency is more of the typical contingencies of project managers where the manager has plans for dealing with scope changes, budget challenges, and deadlines. The only difference is that instead of just looking at the tactical level, the virtual project manager must also examine the strategic implications. For example, if a project misses a deadline, the project manager understands that more resources must be added to that task in order to get it back on track. This is a linear consideration; however, a virtual project manager using complexity will also understand that there are many other factors to consider, such as employee morale, stakeholder perception, upper management concern for more missed deadlines, and budgetary ramifications (Weaver, 2007). A complexity-based approach will not only address the additional resources, but will also leverage a plan that will handle all the ancillary or associated factors that occur with a slip in the project. Veteran project managers know that the day after a project deadline slips the inquisition starts. Rather than become a victim of this kind of backlash, have a contingency to address all those questions and disseminate that information as quickly as possible.

The project contingency should be one part marketing, one part advertising, and one part damage control. Nothing will sidetrack a project manager more than these external time wasters. The more time the project manager is spending answering these additional questions, the less time the project manager will have for the project. All of these ancillary matters are complexity-based fears. Some would see these as reactionary, and perhaps they are to a degree, but these reactions will be more based upon the culture and the organization and less uniform as some people might expect.

In the end, the virtual project manager must utilize complexity in order to address the complex organizational needs that arise during a virtual project. Questions will always arise, so the more that can be addressed up front, the better off the project manager (and the virtual project) will be. Keep in mind that answering unasked questions will save time and help keep people motivated about a project more than skillful recovery when a project encounters problems.

CHAPTER SUMMARY

Virtual projects utilize complexity without ever realizing it. A virtual project must contend with organizational, cultural, and relationship issues at every turn. Virtual projects lack the robustness of face-to-face projects but have all of the same requirements and more. Consider how much additional communication is necessary in order to make up for the fact that people cannot interact naturally together (Duarte & Snyder, 2006).

If nations are fearful of their neighbors because they cannot see them, communicate with them, or interact with them on a regular basis, then imagine the rift that exists in a virtual project. Understanding the complexity of social organizations will help a virtual project manager better address these challenges in advance. The more that can be done up front, the better.

Consider this fact: People often spend weeks, months, or years finding the perfect home. This means looking at new, used, and sometimes dilapidated homes while in search of the perfect place to live. Some of us have difficulty visualizing a home as anything other than what it is and so it requires that home buyers look at many different homes until they find the right fit. If it takes a long time to make a decision when one can see, touch, and experience a home, imagine how difficult it is for a person to understand a project with a team that cannot even be seen most of the time. Furthermore, consider that when the only news that comes from a project is about missing deadlines, this kind of situation will only lead to questions and inquisitions. Thus, a virtual project manager is best served by addressing these complicated matters with contingencies based in complexity.

REFERENCES

Bass, B. (1990). *Bass & Stodgill's handbook of leadership: Theory, research & managerial applications* (3rd ed.). New York: Free Press.

Bennis, W. (1995). The artform of leadership. In T. Wren (Ed.), *The leader's companion: Insights on leadership through the ages* (pp. 377–378). New York: Free Press.

Cooke-Davies, T., Cicmil, S., Crawford, L., & Richardson, K. (2007). We're not in Kansas anymore, Toto: Mapping the strange landscape of complexity theory, and its relationship to project management. *Project Management Journal*, *38*(2), 50–61.

Duarte, D., & Snyder, N. (2006). *Mastering virtual teams: Strategies, tools, and techniques that succeed* (3rd ed.). San Francisco: Jossey-Bass.

Jarvenpaa, S., Shaw, T., & Staples, S. (2004). Toward contextualized theories of trust: The role of trust in global virtual teams. *Information Systems Research*, *15*(3), 250.

Lucey, J. (2008, winter). Why is the failure rate for organizational change so high? *Management Services*, *52*(4), p. 10–18.

Marcinko, R., & Weisman, J. (1997). *Leadership secrets of the rogue warrior: A commandos guide to success*. New York: Simon & Schuster.

Project Management Institute (Ed.). (2008). *A Guide to the Project Management Body of Knowledge—Fourth Edition*. Newtown Square, PA: PMI.

Roebuck, B. (2001). *Improving business communications skills* (3rd ed.). NewYork: Prentice Hall.

Singh, H., & A. Singh (2002). Principles of complexity and chaos theory in project execution: A new approach to management. *Cost Engineering, 44*(12), 23–33.

Weaver, P. (2007). A simple view of complexity in project management. 2007 PMOZ Conference Keynote address.

Chapter 21

Using Complexity to Address a Troubled Project

Figure 21.1 Once a project goes down the wrong road, it is difficult to recover. Sometimes the only way to get a project back is to start from zero and move forward.

TROUBLED PROJECTS

Projects, whether virtual or traditional, become troubled for various reasons. The Standish's Chaos Report (2009) found that approximately 65 percent of IT projects fail. The report also found the top three reasons that projects succeed:

1. User involvement
2. Executive support
3. Clear statement of requirements

While the Standish Chaos Report focused on IT project failures, this information may be extrapolated to other types of projects as well. This may also be extended to virtual projects; however, the trust factor cannot be ignored (Duarte & Snyder, 2006). This will be discussed in more detail in a few paragraphs.

In an organization with a mature project management environment, there should be metrics and periodic assessments of all projects (PMI, 2008). This should prevent most projects from becoming troubled since steps can be taken early to help the project demonstrate symptoms of being behind schedule, customer dissatisfaction, not maintaining scope, quality issues, an inordinate amount of change, financial issues, or other metric established.

In an ideal situation, when a troubled project has a crisis (such as the client sends a letter of a potential litigation; the project is in a severe overrun on a firm fixed bid project), an assessment team should be assembled and the project should be stopped. The assessment team should be senior individuals, one of which should be a project manager with virtual project management experience, who are not associated with the project (see exception below) but work with the project team. The composition of the team depends on the size and type of the project. The project is stopped while the assessment is conducted. This allows for a thorough and focused assessment. Most likely, this will not be possible if this is a client engagement.

In this situation, the assessment team may have to be larger and work around the forward momentum of the project. The assessment team *must* have the full support of senior management. Consideration should be given to having a member of the project team on the assessment team, if that project person can remain unbiased. However, it should not be the project manager of the troubled project. Again, having a person from the project on the assessment team should be carefully evaluated.

The assessment project manager must do the assessment virtually since the project team is distributed in different locations. If possible, it might make sense to have a face-to-face kick-off meeting of the virtual project manager and the team leads. Alternatively, have a videoconference. The assessment will consist of interviews, documentation reviews, triangulation of data, and finally, an approved recovery plan.

The recovery plan is the findings put into a tactical plan. During the recovery, it is a time to micromanage the project back to a steady state. Each person on the project who is in a troubled area is micromanaged until he or she is able to estimate his or her work to within 10 percent of an established metric. A good metric to use and a best practice is a schedule performance index (SPI), which is a metric within Earned Value. The micromanagement can last from two weeks to a couple of months. Remember, no one wants to be micromanaged. The objective of the recovery project manager is to stabilize the virtual project as quickly as possible.

> Practical Tip: When one is micromanaging a project to assist in the recovery, make sure to communicate this situation to the team. Once it is understood that it is temporary, it will be easier to gain support. If the team feels like they will be micromanaged forever, the team might not give their full support. Gaining the support of everyone, or as many people as possible, is critical to recovery.

When the person can estimate within 10 percent of the SPI then the micromanagement for that area can stop. Once all micromanagement stops, that is when rebaselining for the project is done. That is when the recovery project manager can tell senior management the new scope, budget, and schedule. The recovery project manager should also establish clearly defined assessment points for the remainder of the project life cycle to ensure the project stays on track.

In conclusion, organizations need to ensure projects are monitored for health. This will help to minimize the number of troubled projects that culminate in a crisis. Remember, it is a fine line between a normal project with some issues and turbulence and tipping over into the trouble zone, which ends up with a crisis. In a virtual environment, it is even more critical to have more health assessments to ensure the projects are doing well. Since there is no management by walking around, the leadership of the project and the organization do not have the ability to periodically to random "gut" checks.

> Practical Tip: Monitor progress and status of any project you are involved with. By keeping track of the project, the project manager can keep the project from spinning out of control. The failure of a project often occurs because a project manager failed to keep a hand on the pulse of the project.

HOW COMPLEXITY CAN BE APPLIED TO A TROUBLED PROJECT

One example of how complexity can be applied to a troubled project is when the project manager is in a situation where he or she must show results before the project team might be ready. Failure to show results causes a negative impression and stakeholder impatience. Any veteran project manager will agree that stakeholder impatience and haste can often create the necessity for nonsequential activities. Linear project managers might find themselves paralyzed by this need, which may result in the project managers pushing back upon the stakeholders (or in the worst case, the customers). The project manager may throw forth excuses or explanations that do not help the project, but may represent reality. This may cause the stakeholder(s) to express concern about the project, even when there is no real cause for alarm.

Complexity can assist by offering the project manager a more value-driven perspective than a milestone driven (linear) perspective. Project managers can be pushed to resolve and handle issues out of the typical sequence in order to achieve certain milestones that are important at a higher level (Weaver, 2007). This kind of pressure can be exerted upon a project manager in order to achieve certain milestones faster for quicker results. This would be in direct conflict with what is advocated by the Project Management Institute™ (PMI) and most project management methodologies. For example, according to section 6.5 of the *PMBOK*® (PMI, 2008), sequence activities should be Inputs lead to Tools & Techniques, which result in Outputs (Figure 21.2).

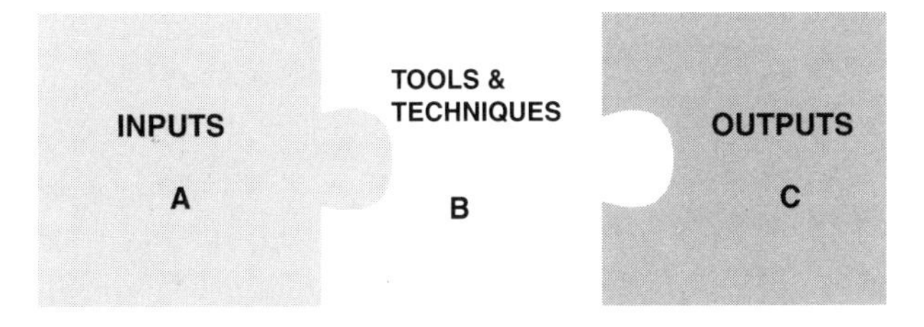

Figure 21.2 Inputs leads to Tools & Techniques, which leads to Outputs.

Too often when stakeholders begin to pressure the project, it can result in what is shown in Figure 21.3.

This example seems to be more expedient; it often results in the application of the incorrect tool or technique. The project manager must resort to intuition or previous similar experience without taking into account the actual project, which can result in more future problems. It can also cause the project team to lose faith in the ability of the project manager and can result in a loss of trust because it appears that the project manager is making decisions in a vacuum. All of these perceptions become reality if no other reality is presented. However, the project manager must make sure to communicate this type of action to the team to the team to avoid any misunderstanding. The project manager may feel he or she is losing some authority by admitting to acquiescing to this type of pressure. This may result in the team trying to circumvent the project manager to reach a higher authority. The openness of communication is often better than the other results.

It is precisely this type of social reality that complexity can help the project manager resolve. By making sure to apply whatever values he or she has for the team to the circumstance, the project manager can better explain to the team and to any other stakeholders what is happening and to also offer the advice if the decision taken is not the best, it can be

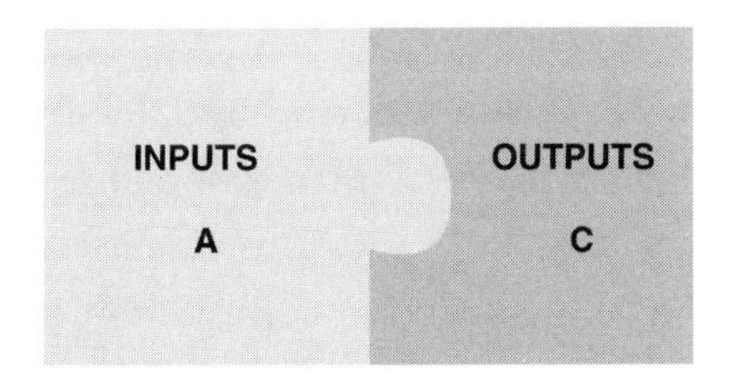

Figure 21.3 Inputs leads to Outputs (Results).

adjusted quickly (Cooke-Davies, Cicmil, Crawford, & Richardson, 2007; Weaver, 2007). Leveraging the network is the best way to address this type of problem in a troubled project. Doing so will assist the project manager with future decisions, and if the project manager finds himself or herself in this type of situation again and the results of the last case were less than stellar, it becomes a way to deflect these problems.

There is no doubt this is a difficult decision for the project manager because it exposes him or her in the short term to potential failure driven by an external force. However, if the project manager is expecting that the team value open and honest communication, then the project manager must be guided by that same value (Duarte & Snyder, 2006). The result of not adhering to the value of open communication is far worse than any smaller issue, which may arise from this type of problem. If one expects to use complexity to assist with this type of systemic and social challenges one must be certain the project manager continues to model and share the values of the project while at the same time remaining a transformational leader.

> Practical Tip: Consider if you are a transformational leader. It is important to learn the skills of being a transformational leader because it will assist you in the application of complexity. The more transformational one is the more that one will be able to leverage complexity.

CELEBRATIONS OF SUCCESSES TO NURTURE FUTURE COMPLEXITY BEHAVIOR

Achievements are the times of great celebration. People strive for great things and the expectation is for a celebration at the successful completion of a task. The celebration should watch the greatness of the achievement. People want recognition for the completion of something extraordinary, for struggling to achieve great goals separately and distinguish themselves from others. Celebration is the closing ceremony of the achievement. It is a gathering to acknowledge the success. Greatness is beautiful for it allows a person's inner light to shine.

What is interesting to understand is that just because one celebrates and recognizes certain achievements, there should also be some recognition to all of the tasks, people, and challenges that achieved the milestones. It should also be a platform for future achievements and to recognize those who will contribute as well as those who will continue to contribute to the success of the project.

The planning of the celebration must be done as a social event. This is where complexity can be interjected into these celebrations. People are social, and offering them the opportunity to plan a celebration can have some even greater benefits (Weaver, 2007). This kind of planning can also be a motivating event. Since people crave a motivating spirit, offering a social outlet for creative behavior and the recognition of organizational values can build the project to a higher status within an organization. It is this kind of complex behavior among people that is often misunderstood. Having flexibility in the planning of these events allows for greater creativity and will encourage more people to want to be a part of the project.

Since anything worth achieving must be earned, goals achieved with little effort are hollow victories. When people put forth great effort they are calling attention to themselves. This kind of complex social behavior can generate unexpected results. When team members are trying to make themselves stand out in a crowd, it helps make the project larger than it is. It allows the project to push out from the huddled masses imprisoned beneath the bell curve of mediocrity. It is this kind of feeling that pushes people to do great things, while others remain satisfied with their more easily achieved goals. Noticing those who aspire and recognizing their greatness is to flatter their soul.

The celebration of success becomes a social release and reward for deeds of greatness. Many of us have seen or visited various monuments to an individual's success, sacrifice, or service to a notable cause. These individuals are placed apart from others for their dedication to a greater cause, and they are often remembered long after their deaths. As much as each of us will probably have exceptional careers and become well regarded and productive members of society, most of us are unlikely to warrant a monument to our achievements. Just as before, the majority of those working on a project will toil in obscurity and receive no long-term recognition for a project that they were a part of; a celebration of success is a fleeting temporal recognition for an otherwise hidden achievement. This then becomes part of the project legacy and becomes an important part of the reputation of an otherwise unknown project team great.

Great goals beget great rewards. Average goals beget average rewards. By embracing complexity, one is deciding to allow individuals to seek more. According to John Wooden (1910–2010), longtime basketball coach of UCLA with ten NCAA titles, and considered one of the most successful college basketball coaches in history: "Success is peace of mind, which is a direct result of self-satisfaction in knowing you made the effort to do the best that you are capable of" (Wooden, 2010). Success comes from doing the best that one is capable of. People need to push themselves towards greater success by taking chances and by learning to always do their best. Allowing people to take the time to do their best makes the project take on more dimensions than others might have believed possible. Greatness is truly deserving of celebration so aim for greatness and celebrate achievement.

> Practical Tip: Making a celebration great makes it an important marker for the future. These types of celebrations become a sort of icon of greatness. Sometimes these events can take on a life of their own as the story is told and retold of the celebration. A great celebration eclipses the end of the project. Learn to make a legendary celebration and you will create a legendary team.

CHAPTER SUMMARY

Troubled projects are a fact of life in business. Every project starts out with the best of intentions and a view toward optimism; there are always internal and external factors that can lead to the ultimate failure of a project. Some projects may have been doomed from the start due to lack of commitment by stakeholders or due to lack of funding. It is important to understand that when a project is in trouble, there should be a cultural drive to try to correct the project in a manner to make it successful. In some cases this may mean walking away from the project or shutting it down. This is probably the hardest decision of a project manager because it often gives the stigma of failure or possibly ineptitude. Regardless, the project manager must always look to making the best decisions with a troubled project, even if the decision might be temporarily poorly received by the organization.

The project manager must learn how to become a transformational leader in order to move the project from where it is to where it needs to be. The more the project manager can transform the team, the more success the project manager can be about moving the project back on track. Understanding that humans are complicated and act and react from social forces can be used to restore morale in a troubled project in order to achieve future success (Pievani & Varchetta, 2005).

Celebration is an important element in troubled projects, because this is something that often is forgotten. Too often people forget that people involved in a struggling project may have been working harder than others, and hard work is often not recognized. A project manager should remember to show appreciation to those working hard, even if the project as a whole may be in trouble. Keeping this perspective will often assist people to keep putting out the additional effort necessary to move the project past the troubled period.

REFERENCES

Cooke-Davies, T., Cicmil, S., Crawford, L., & Richardson, K. (2007). We're not in Kansas anymore, Toto: Mapping the strange landscape of complexity theory, and its relationship to project management. *Project Management Journal*, *38*(2), 50–61.

Duarte, D., & Snyder, N. (2006). *Mastering virtual teams: Strategies, tools, and techniques that succeed* (3rd ed.). San Francisco: Jossey-Bass.

John Wooden Official Website (2010). www.coachwooden.com.

Project Management Institute (Ed.). (2008). *A Guide to the Project Management Body of Knowledge—Fourth Edition*. Newtown Square, PA: PMI.

Pievani, T., & Varchetta, G. (2005). The strategies of uniqueness: Complexity, evolution, and creativity in the new management theories . . . or, in other words, what is the connection between an immune system network and a corporation. *World Futures*, 61. Milan, Italy: Routledge Taylor Francis Group.

Standish Chaos Report. (2009). Standish Group.

Weaver, P. (2007). A simple view of complexity in project management. 2007 PMOZ Conference Keynote address.

Chapter 22

The Future of Complexity

Figure 22.1 As far as the eye can see, there is opportunity. The current path of complexity is not fully written, so just as this bridge goes over this small portion of the everglades, complexity is about the vast unknown that is waiting to be discovered.

WHERE WILL COMPLEXITY BE IN THE FUTURE?

In order to understand where complexity will be in the future, it is important to review where project management has been in the past. When one realizes that large leaps of progress have happened to date, one can better appreciate the future of complexity and project management. The current history of project management has been divided into two distinct periods the historical period of project management, and the modern period of project management. A review of these two periods will make clear that complexity has become the third and newest period of project management.

Historically, project management has been dominated by leaders whose power and will controlled the ancient projects of the past. Since those in power documented evidence of project management throughout the ages, pharaohs, King Solomon, and the Inca Empire leaders were extremely wealthy and powerful (Brier, 2002; 1 Kings 5, King James Version; Hyslop, 1984) and did not demonstrate concern for cost and schedule. The *PMBOK*® (PMI, 2008) states that a project assists an organization to realize its business approach, which normally includes meeting organizational strategies and project goal achievement, including budgets and timeframes (Toney, 2002).

Documentation of ancient engineering, architectural, and military feats may have included project management principles. There are detailed descriptions of the Incas constructing intricate road systems (Hyslop, 1984). Zhuge Liang's book *Art of Management* (Pheng & Lee, 1997) explores the East Asian culture. Detailed accounts of Solomon's temple and palace are found in the Bible (1 Kings 5:1 – 7:51, King James Version; 1 Kings 5:5, King James Version), and Brier (2002) offers accounts on the construction of the Pyramids.

The beginnings of modern project management have been recognized as the late 1950s and early 1960s (Simpson, 1970) with the development of the Project Evaluation and Review Technique (PERT). Leifer, O'Connor, and Rice's (2001) studies find that within technology companies, the successful project manager is a communicator and a negotiator and accomplishes tasks through cooperation and coordination rather than the imposition of power and will upon the constituency. Technology and globalization have altered the traditional PM environment (Montoya-Weiss, Massey, & Song, 2001; Townsend & DeMarie, 1998). Technology has allowed many projects to shift away from the traditional office setting. Technology allows and even encourages businesses to conduct PM in a virtual environment (Leifer, O'Connor, & Rice, 2001; Dess & Rasheed, 1995).

Modern project management is defined as the directing of a group of individuals, each of whom has specific tasks for which they are accountable, share a common mission, and focus upon a common purpose, which must be completed in a specified period of time. The modern project manager utilizes organizational skills, supplemented by technology in order to complete all the necessary tasks in the allotted time. Furthermore, modern project managers are valued for their ability to complete tasks according to the plan and within the specified budget. By contrast, historical project managers were prized for completing the task at hand regardless of budget or consequences.

Throughout history, variations of project managers and project management disciplines have been critical to the completion of ancient and modern-day projects. Modern project management disciplines evolved from the formation of the Project Evaluation and Review Technique (PERT) during the late 1950s and early 1960s (Simpson, 1970). Despite changes in project management disciplines, modern project managers must continue to utilize their organizational skills, supplemented by technology and by project individuals, virtual and nonvirtual, to complete the deliverables of the project in a timely manner within a global marketplace.

However, more recently project managers are emerging with skills and abilities beyond linear thinking. A project manager today cannot expect to manage a project with the same tools as were available in the 1950s or 1960s. Technology has certainly changed, but more importantly the understanding of project management resources has changed in a way that has made complexity part of the process.

WHY CHAOS AND COMPLEXITY ARE HERE TO STAY

Chaos and complexity have only been recognized in the past few decades as a viable field of research and study. In the past, human and scientific study and research has been obsessed with finding a single right answer to any single question. For example, people have been looking for the best leadership style, the most productive organizational type, the best method of communication, when perhaps a better line of research would be to find the best collection of answers to any set of questions. Empirical research has been based upon a one to one relationship between questions and answers. Anything more would appear to need further study until the best single solution can be found. In mathematics, one plus one always results in two. However, human systems never seem to have the same kind of satisfying solution.

When one examines scientific study, there has always been a quest for the single variable universe. For too long, science has tried to reduce everything down to a single variable mathematical expression. One of Einstein's greatest achievements, $e = mc^2$, was an attempt to explain energy in only one possible way the universe. Yet, even this solution is not as tidy as one would think. The reality is that a human system, and the universe, is usually a very messy place where the boundaries are blurred. One only has to look out the window to see that complexity is everywhere.

So, one may ask, how is complexity all around us and how was it missed for so long. The answer is that people have failed to see it because it was not orderly and in many cases society superimposed order onto complexity-based systems. Delving a little deeper uncovers certain basic assumptions and orders not as orderly as they should be. Looking out the window, one may find a number of flying creatures outside. These insects and animals are organized by genus and species, by order and by type. This artificial taxonomy classifies related creatures and the theory of evolution makes for a compelling argument on how these creatures evolved to be different based upon the environment. If the environment was the only variable in the equation, then why don't all flying creatures fly in the same way? If there was a single variable to flight, then shouldn't blue jays, eagles, ducks, hummingbirds, butterflies, dragonflies, bats, and ladybugs all have evolved to fly in the same manner. In this short list, one already sees that some wings have feathers, some do not, and the animal kingdom cannot even agree that two wings are enough (dragonflies have four). This diversity is the expression of complexity.

Perhaps the example about flight is not enough. Check the aisles of the local pharmacy, drug store, or supermarket for different pain relievers. If pain relief is based upon the blocking of the pain receptors in the body and reducing swelling in different areas, would not one type of pain reliever be sufficient for everyone? Instead one finds a whole host of different types and brands. Some might say that all pain relievers are the same; many people that say they are not. Some people have allergies to different types, and others have brand preference.

Even doctors, skilled and trained in prescribing medicines, understand that it often takes a few different medications before finding the one that works best for the individual. If one seeks the scientific reason why someone might take a certain blood pressure medicine and another person uses a different one with similar results (lowering blood pressure), there is not a clear scientific reason other than that some medicines work better or worse for different people. Furthermore, if these same two people with a similar problem were to switch their medications, they would end up with different results. When one digs in the direction of the unknown, that is when one finds complexity. Complexity is not about knowing but about not knowing. About understanding that people and human systems are complex and fluid, these kinds of systems cannot be described by single variable representation. In some cases, a leadership or team model might describe how it works sometimes, but it lacks the 100 percent repeatability and replication necessary to create an immutable law of nature. It turns out there are fewer laws of the universe than expected. The view of the universe is that there should be order and organization to everything, yet there is now reason to believe there are fewer laws and even those laws are subject to revision.

Furthermore, there are still sections of knowledge that are not understood or perhaps are additional areas of complexity. For example, understanding how we learn how to read and speak is another aspect not fully understood. There is scientific information that Broca's area and Wernicke's area of the brain controls speech and language, but there is not a single way that people learn how to read. There is evidence for both whole language and phonics as the best method for learning how to read; it seems that people learn through a combination of these skills. Despite scholars' attempts to identify the one true way of learning language, it seems that a combination of efforts makes proficiency at language a success. It is possible to learn one skill and not the other to be an acceptable reader, but as best as experts can tell, students need a combination of these two skills in order to become successful readers. Hence, the more in depth one examines the surrounding world, the more one sees where complexity has application. People, animals, and natural systems, such as the weather, are all part of the unpredictable universe.

As there is more research in the world regarding complexity, there is more to understand, but there will always be mysteries that are not fully understood and accepting that fact allows us to leverage those complexity-driven systems in order to better handle a project. Understanding that people do not always behave like cogs in a machine, nor should that be the manner to treat them. To some degree there is always a degree of unpredictability in any human system. After all humans do not operate under the 100 percent or 0 percent productivity model like machines; humans can range in productivity from 0–100 percent.

> Practical Tip: Remember that people are not machines and even a top project team member can have a bad day. It is not a matter of giving a person space on a bad day; it is a matter of understanding what drives a person and what helps make them move past that bad day. Consider what communication might be necessary and work toward that interaction to help everyone move closer to 100 percent. Consider the difference between action and interaction.

THE NEXT EVOLUTION OF CHAOS AND COMPLEXITY

Complexity, like anything else in life, has a past, present, and future. The past of complexity is based in chaos theory and as astronomers and scientists continue to tackle the motions of the universe, we find that this theory permeates more than was initially thought. As one considers more about complexity and its application in life, consider the following listing of items to determine if any of the following nine items are representative of complexity.

1. Wine
2. Plants
3. Animals
4. Asphalt
5. Stars
6. Weather
7. Stock market
8. Art
9. People

In order to determine which of these represent complexity, a review of each of these nine items will be done to establish what complexity is and what is not.

A fine wine, made from grapes, can have tastes of grass, smoke, tobacco, fruit, tannins, citrus, and dozens of other taste and smells. Wine offers an excellent example of how something (grapes) can become something so complex that it can offer our senses tastes and smells that would normally never be associated with the component parts (grapes) (Clarke, 1995). Wine is unique among beverages as it can stimulate all of our different scent and taste receptors and offer different sensory experiences. It is no wonder that wine making is considered an art. The winemaker is uniquely enabled to take the product in any direction of their choosing in order to create a familiar yet satisfying product.

Plants are complexity based because one cannot predict with mathematical accuracy when a plant will sprout, will bear fruit, or will flower. Growing plants is not an exact science but merely an approximation based upon prior observations and experience. Farmers are not so much accountants of the land as much as cultivators of what nature brings.

Animals are complexity based because they are living, breathing, thinking creatures that may live by instinct and reaction but are nonetheless actors on their own stage of life. Even domestic animals are not 100 percent predictable in their response to stimuli (Cooke-Davies, Cicmil, Crawford, & Richardson, 2007). Some owners of dogs might believe that their favorite pet is completely loyal and perhaps most are. Living with a cat for a while may truly help a person understand complexity in animals.

Asphalt is a common product in our lives but few people know the composition of the material. It is a petroleum-based product that is basically the leftover material from the refining process mixed with some other elements to make it into the familiar streets that line our cities and towns. It is liquid at higher temperatures, yet it is solid at the normal temperature range of our planet. It may appear that this is just a simple, one-dimensional material but asphalt is also related to the same fuel that powers most of the world's commercial ships. This material is diverse in its uses and is another reflection of complexity.

Stellar motion has long been the realm of complexity. Chaos theory has explained some of the motions of celestial bodies where Newtonian science has fallen short. Even the stars themselves are a complex item because a star is an enormous ball of gas that at a certain size suddenly implodes and self-immolates. Again, science is a little vague on when this situation occurs; if there were an exact size for this to occur, we should find that all stars should start out as the same size, yet somehow this is not the case in the universe.

The weather represents complexity—if only it were not so complex it would make planning time at the beach a little more predictable. Satellites have certainly improved the predictive ability of meteorologists, yet they do not seem to have nailed down exactly what tomorrow will bring.

The weather and the interaction of weather systems is certainly another complexity-based system.

The stock market is example of a complexity-based system. The stock market moves with the feeling of nations, the concerns of business, the impression of people; sometimes it is tied to the weather and natural disasters, and sometimes it changes for no apparent reason at all. It is certainly a system that has so many links and interdependencies that it is unlikely that anyone will be able to understand this system.

Art has been defined by some as anything that elicits an emotional response. Others define art as anything that others prize and will be willing to exchange something for. The ethereal nature of defining art makes it a complexity-based item. Art is complexity for it can evoke emotions, it can cost a fortune, it can be unique, it can be made by one for the pleasure of many, and it is something that has more facets than a diamond. It is also at times a great mystery, and so this multidimensional item should almost be synonymous with complexity.

People are some of the most complicated natural systems on the product, with so many related and interdependent systems that science has no good explanation for why everything interacts with such precision. Humans may have explanations for how these interdependent systems have been able to become so effective together, yet these interactions are still not fully understood. Even the human mind is not fully understood, yet it has been studied for thousands of years. There are different models to understand how electrochemical reactions occur, yet it is not fully understood where the mind even resides. Additionally, human society, which has been under development since the first humans banded together for self-preservation, has long been studied yet no definitive explanation has been forthcoming. Human society is still one of the vast, unexplained arenas of complexity and will likely remain that way forever.

> Practical Tip: Consider how different human actions are in response to the environment. Consider how one interacts with co-workers, as compared to how one interacts with friends. Then compare those to how one would interact with a teacher. Why are these social interactions so different and what drives the different interaction? Is this kind of interaction predictable?

Complexity is everywhere and the clear answer for the question posed earlier is all of the above. Complexity covers every aspect of human life and as one looks closer one will find that complexity is present under every stone and behind every great known and unknown. If one looks closer at the world today, the world is a place of complex interactions where the results are always greater than the sum of the parts.

Moving forward, the only clear reality is to consider this a world of complexity waiting to be understood. Each of us might not totally understand all of the interactions and interdependencies, but we can use the ones that we do understand and the ones that we do see. Ultimately, the goal is to find more complexity so that it can be leveraged to one's advantage in life, in business, and in a project. Keep looking—there is always more complexity out there than one realizes.

> Practical Tip: Complexity is everywhere and the more that a project manager can leverage these interactions and interdependencies, the better they will be able to manage a project.

CHAPTER SUMMARY

Project management continues to evolve and develop. Complexity is part of the world today and the more the project manager can leverage these dependencies and interdependencies, the greater their ability to harness the strength of small groups, large groups, and even communities. Complexity is all around and one just needs to accept that one cannot understand all of these interactions, then move toward understanding the ones that can be comprehended so that those can be used to advantage. Once those few are harnessed, then one needs to continue to expand upon the known to move toward the unknown in order to create a more precise method of handling all sizes of groups.

> Practical Tip: Complexity can be seen in every human interaction and a project manager would benefit from learning as much as possible about these interactions to learn how to better manage a project.

CONCLUSION

Complexity is everywhere and more than ever complexity can assist business in achieving greater results with fewer resources. The more managers can apply complexity, the better they will be able to manage others.

Complexity is about understanding the small in a way that it can be applied to the large. Projects will only become larger and more complex, so the project managers who learn to leverage complexity will be able to handle larger mega-projects or programs (McKinnie, 2007). The Project Management profession, as a whole, needs to recognize that complexity theory is now a force in project management.

The *PMBOK® Guide* (PMI, 2008) is currently lacking information about complexity theory, but there is still an opportunity to include complexity in a future revision of the *PMBOK®* (PMI, 2008). Many competing leadership techniques are available to project managers; however, none offer the same applicability and flexibility as complexity. Each passing month offers a wealth of new resources to project managers, and the sooner complexity is recognized as a world-class practice, the sooner project managers can reap the rewards.

REFERENCES

Brier, B. (2002). The other pyramids. *Archaeology*, *55*(5), 54.

Cooke-Davies, T., Cicmil, S., Crawford, L., & Richardson, K. (2007). We're not in Kansas anymore, Toto: Mapping the strange landscape of complexity theory, and its relationship to project management. *Project Management Journal*, *38*(2), 50–61.

Dess, G., & Rasheed, A. (1995). The new corporate architecture. *Academy of Management Executive*, *9*(3), 7.

Hyslop, J. (1984). *The Inka road system*. Orlando, FL: Academic Press.

Leifer, R., O'Connor, G., & Rice, M. (2001). Implementing radical innovation in mature firms: The role of hubs. *Academy of Management Executive*, *15*(3), 102.

McKinnie, R. (2007). *The application of complexity theory to the field of project management* (UMI No. 3283983).

Montoya-Weiss, M., Massey, A., & Song, M. (2001). Getting it together: Temporal coordination and conflict management in global virtual teams. *Academy of Management Journal*, *44*(6), 1251.

Pheng, L., & Lee, B. (1997). "Managerial grid" and Zhuge Liang's "Art of management": Integration of effective project management. Management Decision, *35*(5/6), 382.

Project Management Institute (Ed.). (2008). *A Guide to the Project Management Body of Knowledge—Fourth Edition*. Newtown Square, PA: PMI.

Simpson, J. (1970). *The Polaris executive.* Public Administration, *48*(4), 379.

Toney, F. (2002). *The superior project organization: Global competency standards and best practices.* New York: Marcel Dekker.

Townsend, A., & Demarie, S. (1998). Keys to effective virtual global teams. *Academy of Management Executive, 12*(3), 17.

Index